Non-Traditional Machining Handbook

by
Carl Sommer

Carl Sommer is president and owner of the largest wire EDM job shop
west of the Mississippi—Reliable EDM Corporation

 *Advance* • HOUSTON
PUBLISHING, INC

Advance Publishing, Inc.
6950 Fulton Street
Houston, TX 77022
Phone (713) 695-0600

www.AdvancePublishing.com

Printed in the United States of America

Library of Congress Catalog Card Number

Sommer, Carl, 1930-
   Non-traditional machining handbook / by Carl Sommer.
     p.   cm.
ISBN 1-57537-325-4
   Includes bibliographical references (p.   ) and index
   1. Machining Handbooks, mauals, etc.   I. Title
TJ1185.S682   2000
671.3'5--dc21   99-20676
                                                         CIP

About the Author

Carl Sommer, owner and president of Reliable EDM Corporation in Houston, Texas, the largest wire EDM job shop west of the Mississippi, is uniquely qualified to write a book on non-traditional machining.

His introduction to the machine tool trade goes back to 1949, at a New York City machine shop. It was not long before Sommer began working as an apprentice tool and die maker where he learned to make dies with hand files and a filing machine.

During the next twenty years, Sommer gained broad, valuable experience in virtually all areas of the field—precision tools and dies, fixtures, and short run production from such companies as IBM, Gyrodyne, Thikol, Fairchild Stratus, Remington, and Sikorsky Helicopter. He operated all the machines, worked in the inspection department, and made precision dies where parts were ground to within .0001″ (.0025 mm). He eventually became a foreman in a tool and die and stamping company.

For most of the 1970s, Sommer taught as a New York City high school teacher. During this time, he also conducted extensive research into the problems facing America's educational institutions. This research, as well as proposed solutions, culminated in writing the book, *Schools in Crisis: Training for Success or Failure?* This book has been highly acclaimed by educators.

Sommer moved to Houston, Texas in 1978, and re-entered the machine tool industry—first as a tool and die maker, then as a tool designer for one of Houston's largest tool and die and stamping shops. He advanced to the position of operations manager, and for more than five years managed the entire company.

In December 1986, Sommer began Reliable EDM with his two sons, Steve and Phil. With the capable leadership of his two sons, within just nine years, Reliable EDM became the largest wire EDM job shop west of the Mississippi. With his son, Steve Sommer, M.E., Sommer wrote Wire EDM Handbook. The handbook, currently in its third edition, has been well received by colleges, trade schools, and industry personnel. This book includes much of the Wire EDM Handbook.

Sommer also has experience with laser cutting and ram EDM. His professional and industry experience has led him to serve on the Texas State Board of Education Review Committee for Essential Knowledge and Skills in the field of Career and Technology Education. He has also taught a Junior Achievement economics summer course to high school and university students at Prague University, Czech Republic.

Other works by Sommer include a twelve-part series of mathematics books promoting adult education, an adult literacy series with an emphasis on phonics (see back of book for information), and eighteen children's picture books encouraging virtues for successful living.

He has been happily married for more than forty years. He has five children and seven grandchildren. Sommer hopes this book will help keep America's businesses abreast of emerging new technologies and help them to remain competitive in the twenty-first century.

Following are pictures of Reliable EDM and some of its equipment. They can be reached at:

Reliable EDM, Inc
6940 Fulton St. • Houston, TX 77022
800-WIRE EDM (800-947-3336)
Tel. 713-692-5454 • Fax 713-692-2466
Web site: www.ReliableEdm.com
Email Phil or Steve@reliableedm.com

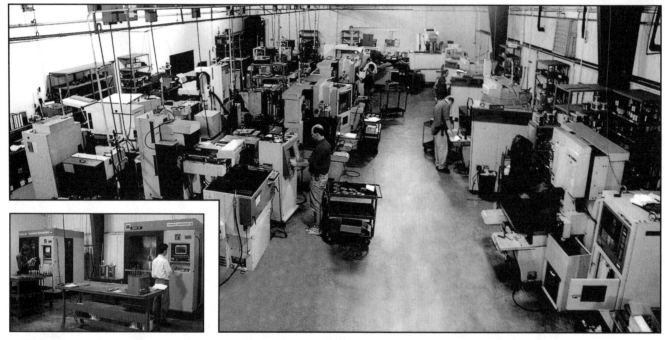

Acknowledgments

I want to thank all those who have written articles about the various non-traditional processes, those who have reviewed the various chapters, and the many manufacturers who have given me permission to use their pictures and provided me with much valuable information for this book. Without their help this book could never have been written.

I have asked experts in each of the various non-traditional machining areas to review the book and offer any suggestions to improve the material. They have been extremely helpful. Since this is an ongoing project of keeping abreast with what is transpiring in the non-traditional machining field, I have been constantly updating the material, so material may have been added or changed since the reviewers have read the material.

My sincerest thanks to the following:

Wire EDM: Ken Goodman—Charmille Technologies, Greg Langenhurst—Mitsubishi, Chris Norman—Sodick

Ram EDM: Zack Medlin—Agie USA, Doug Brown—Charmille Technologies, Dan Zeman—Mitsubishi, Joe Darmstadt—Sodick

Fast Hole EDM Drilling: Clive Greatorex—Belmont Equipment Company, Ken Goodman—Charmille Technologies, John Foster—Current EDM

Abrasive Flow, Thermal Energy Deburring, and Ultrasonic Machining: John Stackhouse—Dynetics Corporation, John Matechen and Lawrence J. Rhoades—Extrudehone

Photo-Chemical Machining: Gary Brubaker—Atotech USA, Fred Grimm—Buckbee-Mears, Dick Shute—Photo Chemical Machining Institute

Electro-Chemical Machining: John Matechen and Lawrence J. Rhoades—Extrudehone, Michael Smith—Raycon/AMCHEM

Plasma and Precision Plasma Cutting: Jim Colt—Hypertherm, Ken Hinderliter—Koike Aronson,

Waterjet and Abrasive Waterjet Machining: Bruce Amundsen—Flow System, Greg Mort—Ingersol-Rand Waterjet

Lasers: Thomas A. Burdel—Bystronic, David A. Belforte—Industry Laser Solutions, Terry Vanderwert—Lumonics/Laserdyne, Doug Wagner—Mitsubishi Laser, Dr. Ronald D. Schaeffer—Resonetics

Rapid Prototyping and Manufacturing: Dr. Paul Jacobs—3D Systems, Sharon Christoperson and Mary Stanley—Stratasys

Dedication

I want to thank my wife, two sons, and daughter for managing
our company, which allowed me the liberty to pursue my publishing goals.

Table of Contents

Ram EDM
Unit 3

Unit 4
Fast Hole EDM Drilling

Unit 5
Abrasive Flow, Thermal Energy Deburring, and Ultrasonic
Machining

Unit 8
Plasma and Precision Plasma Cutting

Unit 9
Waterjet and Abrasive Waterjet Machining

Unit 10
Lasers

Unit 11
Rapid Prototyping and Manufacturing

Unit 14
Building a Successful Business

Unit 15
The Revolutionary Future Non-Traditional Machine

Unit 16
Questions

Index

Unit 1
Fundamentals of Non-Traditional Machining

Notes

1 Fundamentals of Non-Traditional Machining

Understanding the Processes of Non-Traditional Machining

In today's highly competitive world, it is essential to understand the non-traditional machining processes. Because many non-traditional processes have had major advances, every manufacturer needs to learn and understand the many advantages of these latest technologies. For example, wire EDM has increased its speed of cutting up to ten times faster from when it was introduced. Likewise the capacity in workpiece weight and height has also increased. At Reliable EDM, we have machined parts weighing up to 3000 lbs., and EDMed beryllium bars, 37 in. (940 mm) tall. (No machine on the market cuts to this height; we modified a machine to cut 38 in. (965 mm) in order to accomplish this.)

Non-traditional machining is really a misnomer. Due to the rapid advances of technology, many traditional ways of today's machining are performed with the so-called "non-traditional" machines. Manufacturers are realizing dramatic results in achieving excellent finishes, high accuracies, cost reductions, and much shorter delivery times.

I have witnessed, firsthand, the dramatic changes in the machining field. In 1949, I started in a machine shop in Brooklyn, NY. A year later, I became an apprentice in a tool and die shop where we used a filing machine and hand files to make our dies. My next employment was in Long Island City, Queens, where I ground precision form dies to .0001″ (.0025 mm).

When the shop was sold, we became a short run machine shop where I witnessed my first NC machine—a retrofitted Bridgeport vertical milling machine. The machine now would locate holes to be drilled and perform milling operations automatically.

After relocating in Texas, I worked as a tool and die maker and then became the operations manager of one of Houston's largest stamping and tool and die shops. At this shop the stamping dies were milled or ground. When the company purchased a wire EDM machine, it revolutionized our tool and die making. Now the most difficult shapes could be cut accurately into hardened tool steel.

In 1986, with my two sons, I started Reliable EDM. One of the major needs I saw was that individuals needed to be educated concerning the benefits of wire EDM, so I sent information to companies describing the process and the capabilities of wire EDM. Within four years, we became the largest wire EDM job shop in Texas; within nine years, we became the largest wire EDM job shop west of the Mississippi. Figure 1:1 shows a portion of our wire EDM company.

Figure 1:1
Reliable EDM—Largest Wire EDM Job Shop West of the Mississippi

Being a tool and die maker, I saw the great advantages of wire EDM for my trade. The real surprise to me after opening Reliable EDM was the many production jobs that we received from companies that had NC equipment. They discovered that it was more cost effective to have work wire EDMed than to job it out or do it on their own NC equipment.

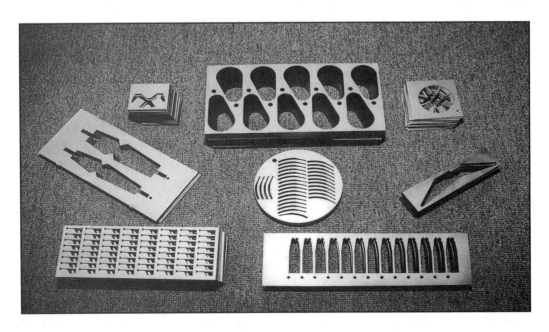

Figure 1:2
Wire EDM Replacing Conventional and NC Machining

The purpose of this book is to educate engineers, designers, business owners, and those making machining decisions to understand and to be able to use the many non-traditional machining methods, and thus make their companies more profitable.

The Machining Revolution

From the earliest ages, individuals learned to use various hand tools to shape objects. As knowledge increased, the use of tools also increased. The industrial revolution arrived and introduced more sophisticated and precise tools such as drill presses, lathes, and milling machines.

Another revolution came—CAD/CAM (computer aided design/computer aided machining). Instead of manually moving machines, computerized programs were downloaded into machines and the operations proceeded automatically. The use of these machines dramatically increased productivity.

With the addition of high speed computers, these machines achieved faster processing times. Then fuzzy logic was introduced. Unlike bilevel logic, which states that a statement is either true or false, fuzzy logic allows a statement to be partially true or false. Machines equipped with fuzzy logic "think" and respond quickly to minute variances in machining conditions. They can then lower or increase power settings according to messages received.

Another great innovation was virtual reality for prototyping and manufacturing. Virtual reality is an artificial environment where any of the human senses such as sight, sound, taste, touch, or smell can be simulated. Virtual reality is simply a computerized experience of real world situations.

In virtual prototyping and manufacturing, a three-dimensional CAD image can be made into a solid part or model directly from the computer. In virtual reality, engineers can visualize and manipulate a 3D model and make any necessary changes before incurring hard tooling costs. This process of rapid prototyping dramatically reduces time and costs.

Other innovations include automatic tool changers, robots, workpiece and pallet changers, and artificial intelligence that enables machines to "think" through complex machining sequences.

The Two Phases of the Computer Revolution

A. Computer-controlled machines, such as lathes, mills, and machining centers, using traditional machining methods which use hard tooling.

B. Computer-controlled machines using non-traditional machining methods which use processes other than hard tooling like wire and ram EDM, abrasive flow and ultrasonic machining, photochemical and electrochemical machining, plasma

and precision plasma cutting, waterjet and abrasive waterjet machining, lasers, and
rapid prototyping. See Figure 1:3

Courtesy Mitsubishi

Wire EDM

Courtesy Agie

Ram EDM

Courtesy Charmilles

Fast Hole EDM Drilling

Courtesy Extrudehone

Ultrasonic Machining

Courtesy Extrudehone

Abrasive Flow Machining

Figure 1:3

Courtesy Buckbee-Meers

Photochemical Machining

Courtesy Raycon/AMCHEM

Electrochemical Machining

Courtesy Hypertherm

Precision Plasma Cutting

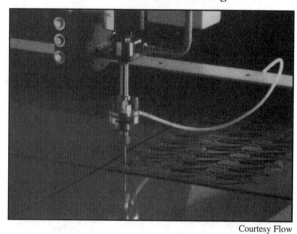

Courtesy Flow

Abrasive Waterjet Machining

Courtesy Stratasys

Rapid Prototyping

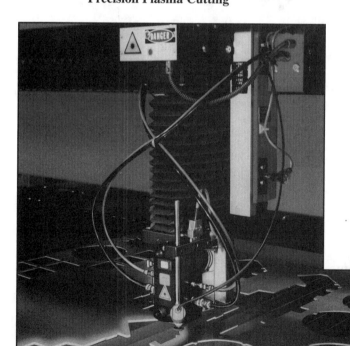

Laser Cutting   Courtesy Bystronics

Figure 1:3

Six Basic Processes to Alter Material in Non-Traditional Machining

There are basically six processes used in non-traditional machining to alter material: electricity, water, abrasives, chemicals, plasma, and light. Understanding these processes will help decision makers to evaluate the opportunities and limitations of these operations.

A. Electricity

Wire EDM and plunge EDM uses electricity to cut electrically conductive material by means of spark erosion.

B. Water

Water jet machining uses high pressure water (up to three times the speed of sound) to cut material. Since water is used, any material can be cut if it is soft enough for high pressure water to penetrate.

C. Abrasives

Abrasive waterjet adds garnet, a silicate abrasive material, to a high pressure water jet. This allows any material to be cut.

Abrasive flow machining is used to deburr and polish parts by forcing a semisolid abrasive media through workpieces.

Ultrasonic machining uses fine, water-based abrasive slurry to machine parts. The vibrating machine causes abrasives to remove material.

D. Chemicals

Photochemical machining relies on chemicals to remove exposed material to etch or cut parts.

Electrochemical machining combines chemicals and electricity to remove material through a deplating process. As a salt solution electrolyte surrounds an electrode, an electrical current passes from the electrode to the workpiece removing the material.

E. Plasma

Plasma and precision plasma cutting systems utilizes ionized gas to cut electrically conductive materials.

F. Light

Lasers rely on highly magnified light to alter materials. Since light is used, both electrical and non-electrical material can be processed. Beside cutting, lasers are used for welding, cladding, alloying, heat treating, marking, and drilling.

The six processes to alter material in non-traditional machining are electricity, water, abrasives, chemicals, ionized gas, and light. See Figure 1:4.

Electricity	Wire EDM and Ram EDM
Water	Water Jet
Abrasives	Abrasive Water Jet, Abrasive Flow, and Ultrasonic Machining
Chemicals	Electrochemical Machining and Photochemical Machining
Ionized Gas	Plasma and High Definition Plasma Cutting
Light	Lasers and some Rapid Prototyping*

Figure 1:4 The Six Processes in Non-traditional Machining

(*Some rapid prototyping systems do not use lasers. Some extrude heated thermoplastic and another system uses photopolymer that hardens when exposed.)

Speed and Accuracy

The speed and accuracy of these various methods vary substantially. An understanding of the characteristics of the various systems can greatly affect the profitability of manufacturers. For example, sheet metal parts can be cut very fast with lasers, but lasers cut parts one at a time. With wire EDM, thin sheet metal parts can be stacked, and the parts will be cut burr free, more economically, and with greater accuracy.

Understanding Accuracy

One of the biggest difficulties in the machining trade is determining required part accuracies. Certain jobs require extremely close tolerances; but excessively close tolerances are often unnecessary and add substantial costs to the machining process. Understanding tolerances is an important asset in reducing machining costs.

To better understand the accuracy of some of these machines, the thickness of a human hair is slightly over .002″ (.051 mm). Some machines can cut to +/- .0001″ (.0025 mm) and closer, one-tenth the thickness of a human hair!

Many manufacturers misunderstand close tolerance measurements. They put on their prints +/- .0005″ (.0127 mm) whether the size is 2 inches (51 mm) or 10 inches (254 mm). For example: the coefficient of expansion of steel is 6.3 millionths (.0000063) per inch (.00016 mm) per degree F. (.56 C). If the temperature of a 10 inch (254 mm) piece of steel rises only 10 degrees F. (5.6 C.), it will expand .00063″ (.016 mm). The 10 inch part could become out of tolerance just from the heat applied by physically handling the steel through heat expansion. See Figure 1:5.

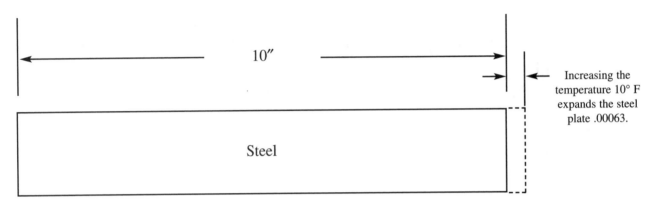

Increasing the temperature 10° F expands the steel plate .00063.

Figure 1:5
Heat Expansion

The Future

For many years, I worked for others. Now as a business owner, I appreciate how the free enterprise system is such an excellent economic way of conducting business. In chapter 29 I share some of my experiences in building a business.

Companies can only remain successful if they actively endeavor to keep their operations competitive. The moment they become careless, others will arise and work more competitively. Eventually, due to their higher machining costs, these inefficient companies will be put out of business. The benefit of the free enterprise system is that surviving companies are the ones that produce the best products at the lowest prices.

Therefore, to remain successful, companies need to keep informed of the newest technologies in order to remain competitive. In addition, they need to train their employees to work efficiently and accurately.

Engineering schools should be concerned that their graduating engineers are properly equipped to enter the workforce knowing the latest technologies. Generally, non-traditional machines are expensive, and schools are usually unable to justify having such machines. This book aims to encourage and educate upcoming engineers and those in management to understand and be able to use the high-tech industrial methods and help our nation to remain competitive.

The future belongs to those keeping abreast and applying the latest technologies to stay competitive. Those becoming satisfied and refusing to look at the new ways of machining, will fall to competitive forces.

Undoubtedly, anyone who has read this far is interested in learning the new non-traditional technologies. For such readers, this book examines the many exciting methods that are available in today's market so they can make intelligent decisions concerning efficiency and competitiveness.

Unit 2

Wire EDM

Notes

2 Fundamentals of Wire EDM

Revolutionizing Machining

Wire Electrical Discharge Machining (EDM), as shown in Figure 2:1, is one of the greatest innovations affecting the tooling and machining industry. This process has brought to industry dramatic improvements in accuracy, quality, productivity, and earnings.

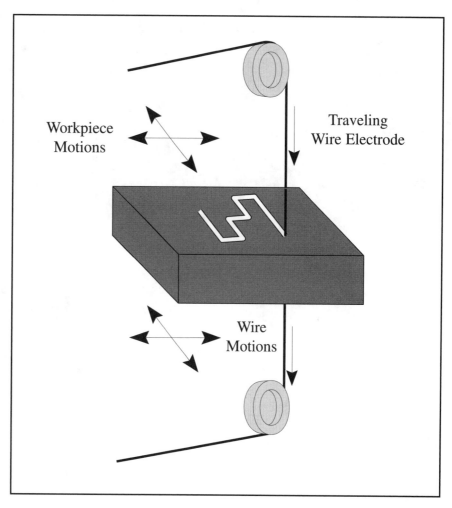

Figure 2:1
Wire Electrical Discharge Machining

Before wire EDM, costly processes were often used to produce finished parts. Now with the aid of a computer and wire EDM machines, extremely complicated shapes can be cut automatically, precisely, and economically, even in materials as hard as carbide.

Wire EDM Beginnings

In 1969, the Swiss firm Agie produced the world's first wire EDM machine. Typically, these first machines in the early '70s were extremely slow, cutting about 2 square inches an hour (21 mm²/min.). Their speeds went up in the early '80s to 6 square inches an hour (64 mm²/min.). Today, machines are equipped with automatic wire threading and can cut nearly 30 square inches an hour (322mm²/min.).

Courtesy Charmilles Technology

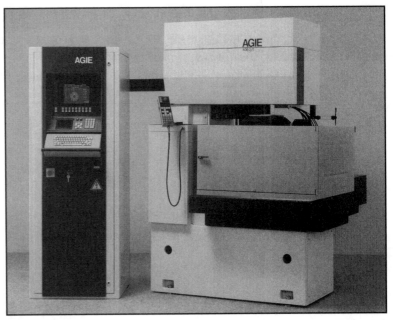

Courtesy Agie

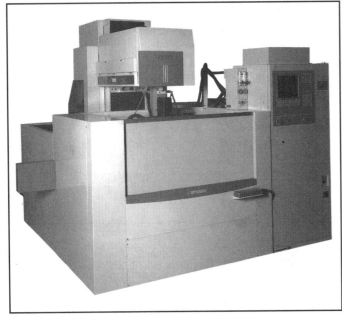

Courtesy Mitsubishi

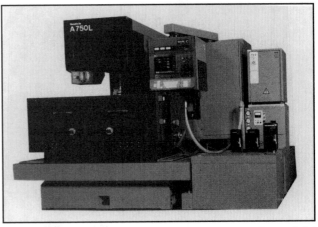

Courtesy Sodick

Figure 2:2
High Precision Wire Electrical Discharge Machines

Some machines cut to accuracies of up to +/- .0001″ (.0025 mm), producing surface finishes to .037 Ra μm and lower. (See Figure 2:3). One wire EDM machine on the market uses a special cutting fluid to produce a mirror-like finish.

Figure 2:3
Wire EDM Skim Cutting

Wire EDM a Serious Contender With Conventional Machining

A brochure published by the Society of Manufacturing Engineers states, "EDM can no longer be considered non traditional machining. With improvements such as CNC, advanced power supplies, ability to handle larger work pieces, and advances in wire, EDM is now commonly used in everyday operations. These improvements are allowing the EDM user to obtain faster metal removal rates yet maintain super-smooth finishes with repetitive accuracy."

Wire EDM competes seriously with such conventional machining as milling, broaching, grinding, stamping, and fine-blanking. Conventional wisdom suggests that wire EDM is only competitive when dealing with expensive and difficult-to-machine parts. But this is not the case. Wire EDM is often used with simple shapes and easily machined materials. Our company, Reliable EDM, receives much work that could be machined by conventional methods. Although many of the customers have conventional CNC machines, they send their work to us to be EDMed.

A large wire EDM company reports that production runs up to 30,000 pieces take 65% of their cutting time. One particular job of theirs would have required fine blank tooling and a 10 to 12-week wait, but EDM was able to finish the hardened, .062″ (1.57 mm) thick stainless steel parts burr-free and on time for their production schedule.

Whether cutting soft aluminum, hot rolled steel, super alloys, or tungsten carbide, manufacturers are discovering it is less expensive and they receive higher quality with today's high-speed wire EDM machines. See Figure 2:4 for various production parts that were cut at Reliable EDM.

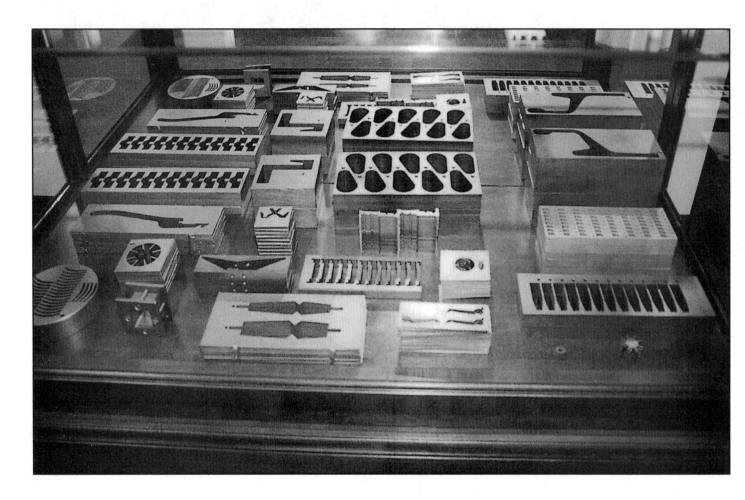

Figure 2:4
Production Wire EDM

New Demands by Design Engineers

As more design engineers discover the many advantages of wire EDM, they are incorporating new designs into their drawings. It therefore becomes important for contract shops to understand wire EDM so they can properly quote on these new designs requiring EDM.

Increasingly, today's drawings are calling for tighter tolerances, shapes that only can be efficiently machined with wire EDM, and alloys difficult to machine, as illustrated in Figures 2:5 and 2:6. With wire EDM these exotic alloys can be machined just as easily as mild steel. When wire EDM manufacturers select the optimum steel to demonstrate the capability of their machines, their choice is not mild steel, but hardened D2, a high-chrome, high-carbon tool steel.

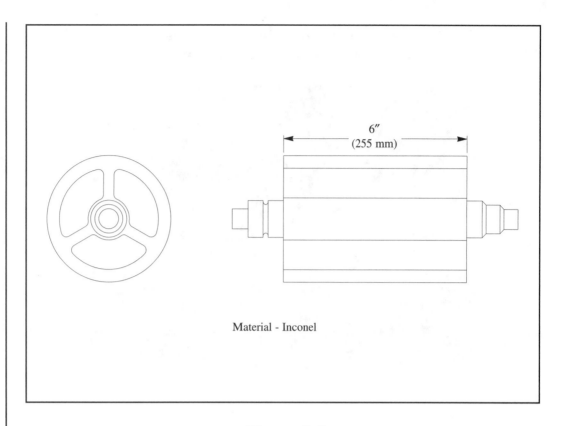

6″
(255 mm)

Material - Inconel

Figure 2:5
Cavities to be Wire EDMed in Inconel

Figure 2:6
Some Difficult-to-Machine Shapes Done at Reliable EDM

Whether cutting in the air or making a full cut as in Figures 2:7 and 2:8, wire EDM has proven to be one of the greatest machining revolutions.

Figure 2:7
Cutting in the Air

Figure 2:8
Making a Full Cut

Fully Automated Wire EDMs

For total unattended operation, some wire EDM machines are equipped with automatic wire threading and robotized palletization. These machines are well equipped to do high production runs.

One company making standard and made-to-order punch and die sets for turret punch presses uses ten wire EDM machines fed by a robot. The robot moves on a track between the two rows of wire EDM machines. After the parts are EDMed, a non-contact video inspection system, interfaced with a computer system automatically examines the work.

General Electric uses 36 wire EDM machines to cut steam turbine bucket roots. Previously, GE used as many as 27 different operations, many of them milling; now it can cut the entire bucket periphery in one pass. Prior delivery with conventional methods required 12 weeks; wire EDM reduced the delivery to 2-4 weeks.

How Wire EDM Works

Wire EDM uses a traveling wire electrode that passes through the workpiece. The wire is monitored precisely by a computer-numerically controlled (CNC) system. See Figure 2:9.

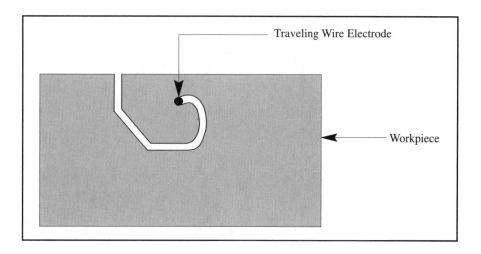

Figure 2:9
Wire EDM
This EDM process uses a wire electrode to remove material.

Like any other machining tool, wire EDM removes metal; but wire EDM removes metal with electricity by means of spark erosion.

Rapid DC electrical pulses are generated between the wire electrode and the workpiece. Between the wire and the workpiece is a shield of deionized water, called the dielectric. Pure water is an insulator, but tap water usually contains minerals that causes the water to be too conductive for wire EDM. To control the water conductivity, the water goes through a resin tank to remove much of its conductive elements—this is called deionized water. As the machine cuts, the conductivity of the water tends to rise, and a pump automatically forces the water through the resin tank when the conductivity of the water is too high.

When sufficient voltage is applied, the fluid ionizes. Then a controlled spark precisely erodes a small section of the workpiece, causing it to melt and vaporize. These electrical pulses are repeated thousands of times per second. The pressurized cooling fluid, the dielectric, cools the vaporized metal and forces the resolidified eroded particles from the gap.

The dielectric fluid goes through a filter which removes the suspended solids. To maintain machine accuracy, the dielectric fluid flows through a chiller to keep the liquid at a constant temperature. See Figure 2:10.

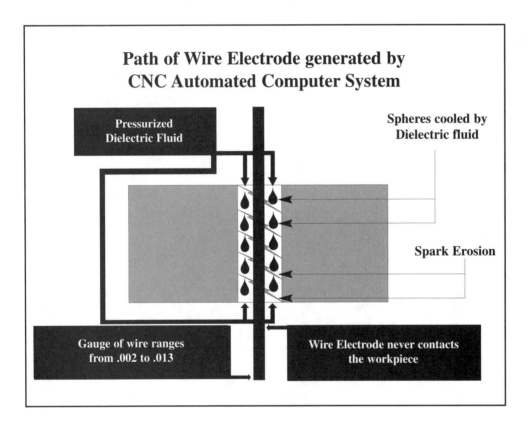

Figure 2:10
How Wire EDM Works
Precisely controlled sparks erode the metal using deionized water.
Pressurized water removes the eroded material.

A DC or AC servo system maintains a gap from .001 to .002 between the wire electrode and the workpiece. The servo mechanism prevents the wire electrode from shorting out against the workpiece and advances the machine as it cuts the desired shape.

The wire electrode is usually a spool of brass, or brass and zinc wire from .002″ (.05 mm) to .013″ (.33 mm) thick. Sometimes molybdenum wire is used. Some machines can even cut with .001″ wire (.025mm). To cut with such thin wires, tungsten is used. New wire is constantly used which accounts for the extreme accuracy of wire EDM.

The Step by Step EDM Process

A. Power Supply Generates Volts and Amps

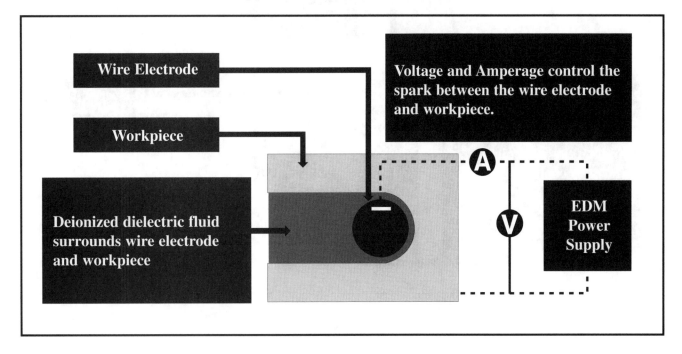

Figure 2:11

Deionized water surrounds the wire electrode as the power supply generates volts and amps to produce the spark.

B. During On Time Controlled Spark Erodes Material

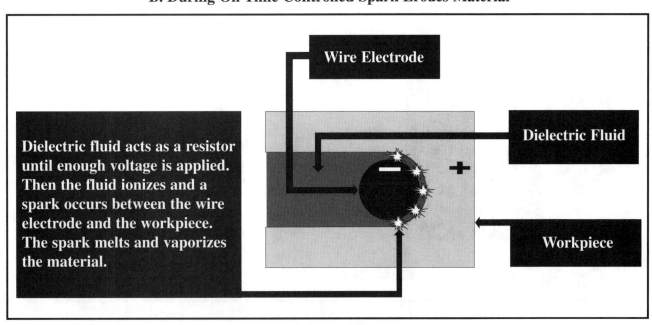

Figure 2:12

The generated spark precisely melts and vaporizes the material.

C. Off Time Allows Fluid to Remove Eroded Particles

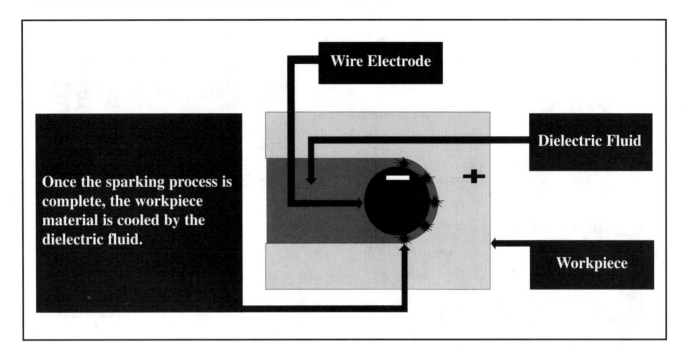

Figure 2:13
During the off cycle, the pressurized dielectric fluid immediately
cools the material and flushes the eroded particles.

D. Filter Removes Chips While the Cycle is Repeated

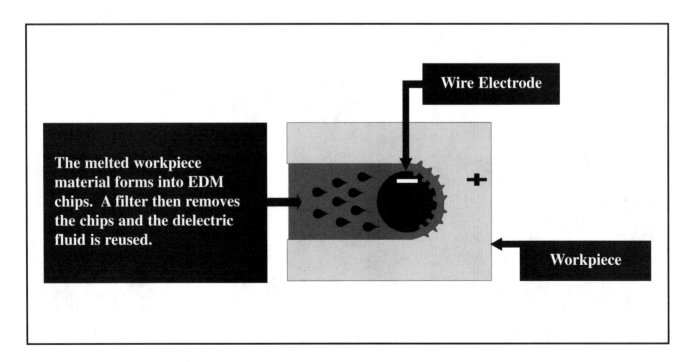

Figure 2:14
New wire is constantly fed, while the eroded particles are removed and separated by a filter system.

Super Precision Band Saw

To better understand the wire EDM process, visualize the wire EDM machine as a super precision band saw with accuracies to +/-.0001 ″ (.0025 mm). See Figure 2:15

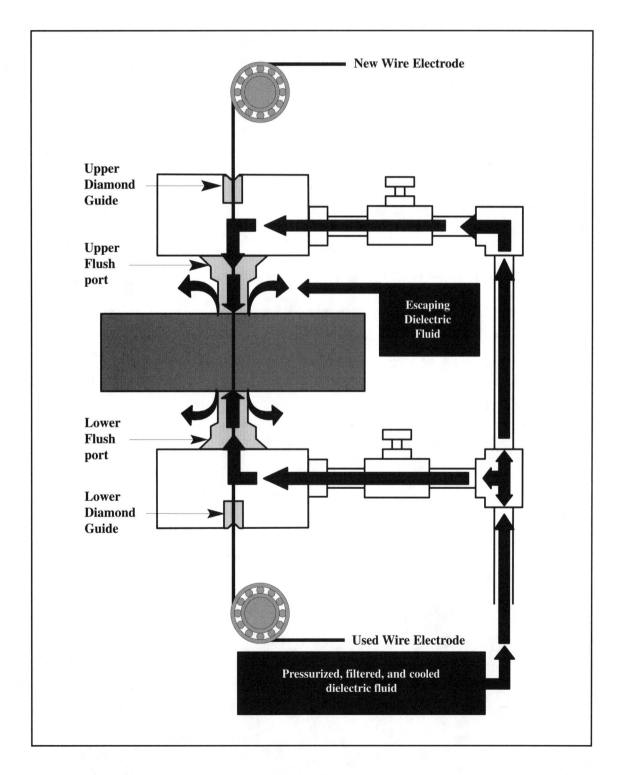

Figure 2:15
A super precision band saw capable of cutting hardened material to +/- .0001″ (.0025 mm).

Three Types of Wire EDM

A. Two Axis

Two axis wire EDM permits only right angle cuts. See Figure 2:16.

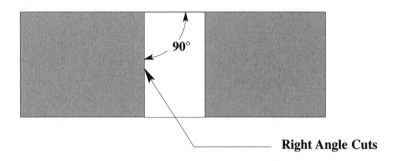

Figure 2:16
Two Axis—For right angle cuts only.

B. Simultaneous Four Axis

Simultaneous four axis machines can produce a taper that is the same shape both on the top and on the bottom, as pictured in Figure 2:17. They are also capable of going from a straight cut to a taper. This type of machine is particularly useful for making stamping dies. A straight land with a taper can also be produced with a secondary skim cut.

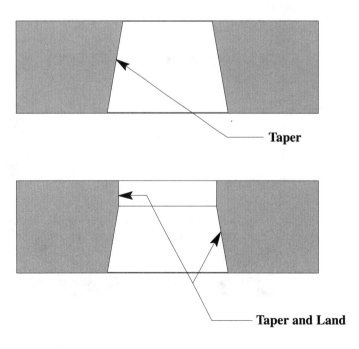

Figure 2:17
Simultaneous Four Axis—For taper and straight cuts.

C. Independent Four Axis

Independent four axis machines can cut a top profile different from the bottom profile. See Figure 2:18. This is particularly useful for extrusion molds and flow valves.

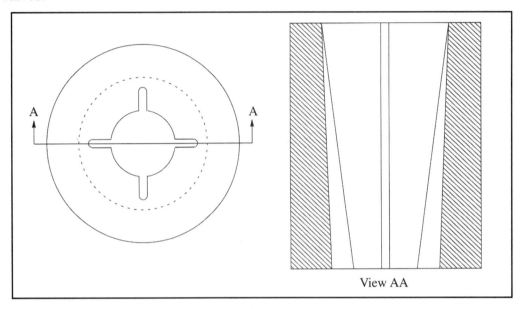

Figure 2:18
Independent Four Axis
Different shapes can be produced on top and bottom of a workpiece.

Parts as shown in Figure 2:19 were produced with independent four axis wire EDM.

Figure 2:19
Independent Four Axis Parts

A computer image of the numbers one and two combined into a single piece is shown in Figure 2:20. (See the second image on the left in Figure 2:19 for the number one and two.)

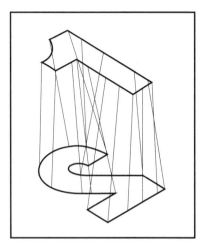

Figure 2:20
Programmed Number One and Number Two.

A picture of the Statue of Liberty combined with a cross is shown in Figure 2:21, and the computer image in Figure 2:22.

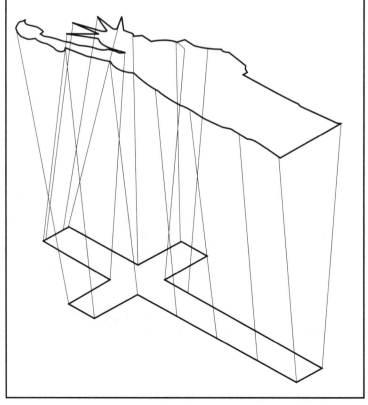

Figure 2:21
Statue of Liberty and Cross

Figure 2:22
Computer Image of the Statue of Liberty and Cross.

Understanding Independent Four Axis

Manufacturers have discovered unique ways of using the capabilities of the independent four axis: extrusion molds, flow openings, injection molds, and many other complex shapes.

To better understand independent four axis, a person can hold a straight wire and move the top and bottom independently. Virtually any conceivable shape can be created within the confines of the travel of the U and V axes. Machines are now capable of cutting 15.75″ (400 mm) with independent angles up to 45 degrees. See Figure 2:23.

Figure 2:23
Wire EDM Machines are Capable of Cutting up to 15-3/4″
(400 mm) with Independent Angles up to 45°.

3 Profiting With Wire EDM

Users of Wire EDM

Parts made with the wire EDM process are used in various fields and industries, such as: medicine, chemical, electronics, oil and gas, die and mold, fabrication, construction, automotive, aeronautics, space—virtually any place where metals are machined.

Benefits Of Wire EDM

A. Production Runs

Because of the new generation of high-speed wire EDM machines, manufacturers increasingly are discovering that wire EDM produces many parts more economically than conventional machining. See Figure 3:1. An additional benefit with wire EDM is that close tolerances can be held without additional cost and without burrs.

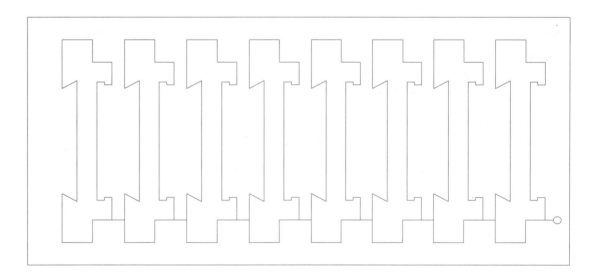

Figure 3:1

**Production EDM—Today's high speed wire EDM machines can produce
many parts more economically than with conventional machining.**

B. Various Shapes and Sizes

With this new technology, any contour (Figure 3:2) and varying tapers can be machined precisely. Extremely thin sections can be made because the wire electrode never contacts the material being cut. EDM is a non-contact, force-free, metal-removing process which eliminates cutting stress and resultant mechanical distortion.

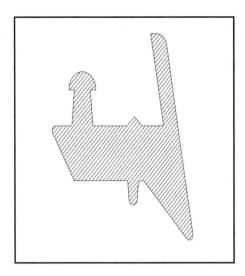

Figure 3:2

Infinite Shapes and Sizes—The most intricate shapes can be machined precisely.

C. Accuracy and Finishes

The wire path is controlled by a CNC computer-generated program with part accuracies up to +/- .0001″ (.0025 mm), and some machines achieve surface finishes well below .037 Ra μm. Dowel holes can be produced with wire EDM to be either press or slip fit.

D. Eliminates Extra Machining Processes

The extremely fine finish from the standard wire EDM process often eliminates the need for grinding or other finishing procedures. When using wire EDM, one should not hesitate to add small radii to eliminate a secondary operation, such as deburring of edges (Figure 3:3). The cost is unaffected by adding radii.

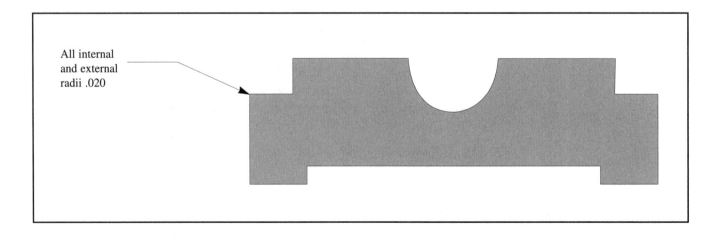

Figure 3:3

Eliminates Extra Machining Process

E. Burr Free and Perfectly Straight Machining

Stamped materials have rollover edges and tapers. Wire cut materials are totally burr free, smooth and straight. See Figure 3:4.

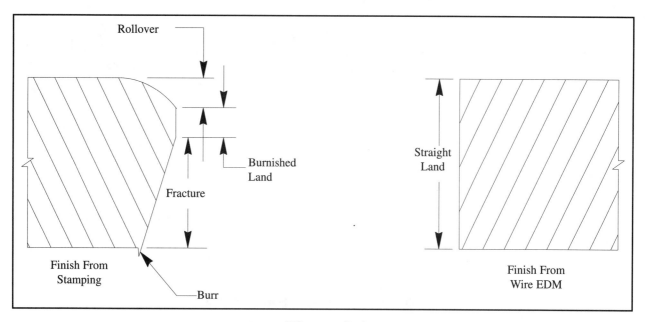

Figure 3:4

Wire EDM Parts are Straight and Burr Free

F. Damaged Parts Can Be Repaired with Inserts

EDM allows a damaged die, mold, or machine part to be repaired with an insert rather than having the part to be remade. An insert can be EDMed and held with a screw, or a tapered insert can be produced so that it can be forced to fit. See examples in Figures 3:5 and 3:6.

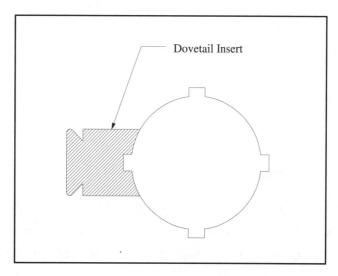

Figure 3:5

Damaged Die Repaired With Insert—Dovetail can be pressed fit or held with a screw

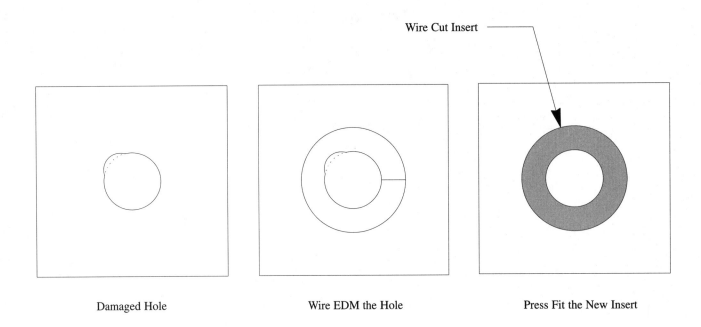

Wire Cut Insert

Damaged Hole   Wire EDM the Hole   Press Fit the New Insert

Figure 3:6

Repairing a Damaged Hardened Hole

G. Less Need for Skilled Craftspersons

Because wire EDM often eliminates extreme precision and time consuming machining processes, it reduces the need for skilled craftspersons. This frees such professionals for more productive and profitable work.

H. Material Hardness Not a Factor

Wire EDM's cutting ability is unaffected by workpiece hardness. In fact, it cuts hardened D2 faster than cold roll steel. The advantage of cutting materials in the hardened state is that it eliminates the risk of distortion created when the material needs to be heat treated.

EDM introduces into the material little heat, and the small amount of heat that is generated is quickly removed by the dielectric fluid. At Reliable EDM we have EDMed hundreds of hardened stamping dies from various tool steels with no negative results.

I. Computers Can Perform Calculations

Since computers program the path for wire EDM, usually only basic math dimensions are needed. Also when exact chord positions on blending radii are required, computer programs can automatically calculate the blending points. See Figure 3:7

Figure 3:7

Computers are used to program the wire EDM path

J. Digitizing is Possible

It is not always necessary to have exact dimensions of a drawing or of a part. By means of digitizing, a program can be made directly from a drawing or from a previous-produced part.

Figure 3:8

Example of a Digitized Drawing

K. Miniaturization of Parts

Wire EDM can machine thin webs with extreme precision, and close inside and outside radii with very fine micro finishes (Figure 3:9). Some machines can cut with wire as thin as .0012″ (.030 mm) wire.

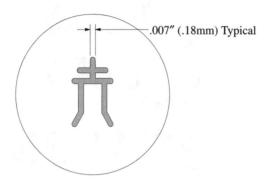

.007″ (.18mm) Typical

Figure 3:9

Miniaturization—Wire EDM can produce very thin webs.

L. Air Machining is Possible

Parts can be EDMed, as shown in Figure 3:10, even when flush ports are not directly against the part. This air machining is a slower cutting process due to poor flushing conditions, but in many cases it is still economical.

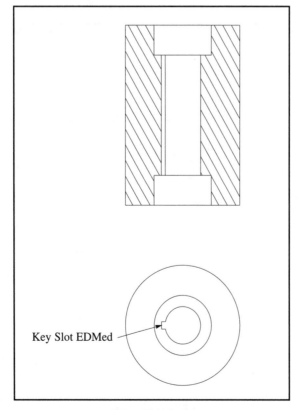

Key Slot EDMed

Figure 3:10

Air Machining—Keyway can be EDMed even though flush ports do not contact the part.

M. Reliable Repeatability

The reliability of wire EDM is one of its great advantages. Because the programs are computer generated, and the electrode, which is used only once, is being fed constantly from a spool, the last part is identical to the first one. The cutter wear of conventional machining does not exist with wire EDM. In addition, tighter machining tolerances can be maintained without additional costs. See Figure 3:11.

Figure 3:11

Quality Work—Display of wire EDM parts at Reliable EDM

Parts for Wire EDM

A. Precision Gauges and Templates

Computer programs rather than costly grinding procedures can produce precision gauges and templates as in shown in Figure 3:12. Since gauges and templates are often thin, making two or more at the same time adds little to the cost of their production.

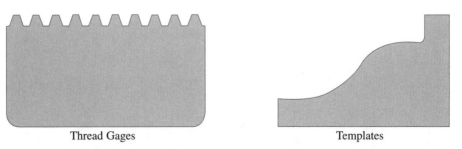

Thread Gages   Templates

Figure 3:12

Precision Gauges and Templates

B. Keyways vs. Broaching

Wire EDM easily cuts precision keyways as shown in Figure 3:13. It also produces hexes, splines, and other shapes, without the need to make special broaches, even from the material in the hardened state.

Figure 3:13

Precision Keyways

C. Shaft Slots

Rather than make a costly setup to machine a slot in a shaft, a simple setup can be made on a wire EDM machine. In addition to saving time, EDM produces no burrs in the threaded area. See Figure 3:14.

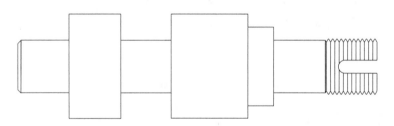

Figure 3:14

Burr-Free Slot in Threaded Area

D. Collets

Conventional machining often distorts collets. If the collets are heat treated after machining, they often distort even more. In contrast, wire EDM can machine collets in the hardened condition and without any cutting pressure, as shown in Figure 3:15

Hole Wire EDMed for Strength

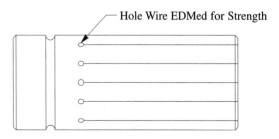

Figure 3:15

Collets can be cut in the hardened condition without any cutting pressure.

E. Parting Tubes and Shafts

Because of the small gap produced by wire EDM—a .012″ (.3 mm) wire produces a .015″ (.38 mm) gap—tubes, shafts, and bearing cages can be parted after machining is completed, as pictured in Figure 3:16.

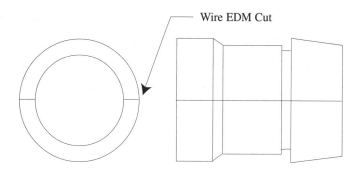

Wire EDM Cut

Figure 3:16
Splitting Tubes

F. Shaft Pockets

Any shaped pocket which goes through a shaft can be machined with wire EDM. See Figure 3:17.

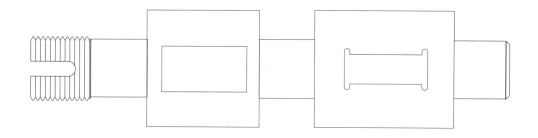

Figure 3:17
Shaft Cutouts

G. Fabrication of Graphite Electrodes for Ram EDM

Graphite electrodes for ram EDM can be machined with wire EDM. One of the great advantages for this is that wire EDM produces identical electrodes.

The cost of producing graphite electrodes is largely determined by the cutting speed of the wire. The cutting speed of various grades of graphites are vastly different. For example, the graphite Poco Angstrofine, EDM-AF5, cuts nearly twice as fast as most of the other grades, EDM-1, EDM-3, EDM-100, or EDM 200.

H. Punches and Dies From One Piece of Tool Steel

With wire EDM, dies no longer have to be made by the costly method of being sectioned and precision-ground. Now the most elaborate contours can be made from one solid piece of hardened tool steel, as shown in Figure 3:18. The reason dies can be cut from one piece of hardened tool steel is due to the capability of wire EDM to cut various tapers. The material thickness of the stamped part determines the taper. After the punch is cut, land on the die section can also be EDMed.

The one-piece tool steel results in a stronger, non-breathing die at a fraction of the cost of a sectioned die. Also, compound dies can be wire EDMed from one piece of tool steel. For detailed instructions on EDMing and fabrication of these low-cost, high-performance one-piece dies, see the book, *Wire EDM Handbook**.

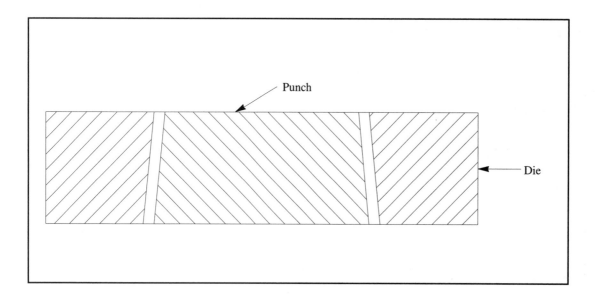

Figure 3:18

One Piece Punch and Die

I. Short Run Stampings

Instead of expensive tooling being produced for short runs, precision parts can be produced with wire EDM. See Figure 3:19. When alterations are needed, they can be made at practically no cost; while alterations with hard tooling are usually costly. Wire EDM can also produce all sorts of special shapes and in various thicknesses.

**Wire EDM Handbook* can be purchased from various sources, or directly from Advance Publishing, Inc. (713) 692-0600. Note: Much of the material in this book is a duplication of these chapters on wire EDM. However, there is in this book much helpful information for EDMing stick punches and for building high-performance one-piece stamping dies.

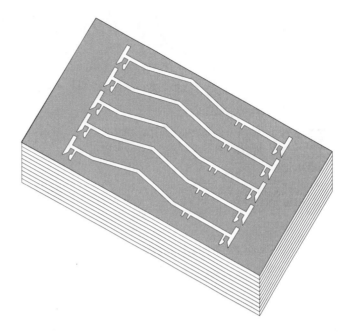

Figure 3:19

Stacked Material to Produce Intricate Parts

J. Molds

Elaborate extrusion molds, with or without taper can be produced economically. See Figure 3:20.

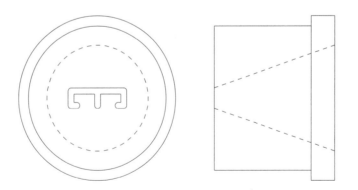

Figure 3:20

Tapered Extrusion Mold

K. Special and Production Tool Cutters

Wire EDM can produce special one-of-a-kind tooling with various tapers, including carbide. See Figure 3:21. When production tool cutters need to be made, they should all be the same to eliminate costly setups and checking procedures when changing cutters. Since wire EDM repeats accurately, this process produces identical production tool cutters.

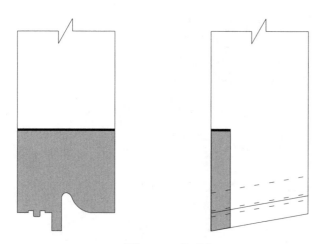

Figure 3:21

Special Carbide Form Tool

L. Difficult-to-Machine Shapes

Wire EDM has dramatically reduced costs for many manufactured parts. Instead of using costly setups and complicated machining procedures to produce parts, wire EDM is often more cost effective. See Figure 3:22 for a difficult production machining operation that could be produced more economically with wire EDM.

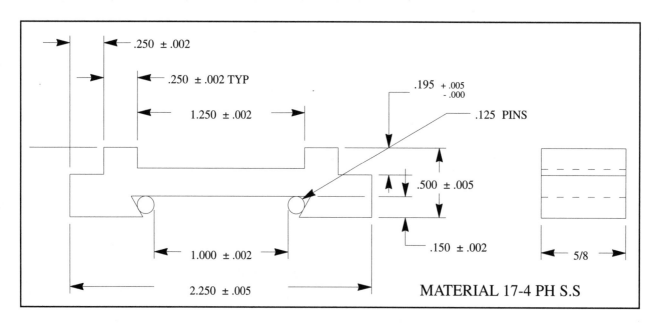

Figure 3:22

Difficult-to-Machine Shapes

M. Other Cost-Reducing Parts

Many other parts can be also economically produced with wire EDM. Following are some samples. See Figure 3:22-28.

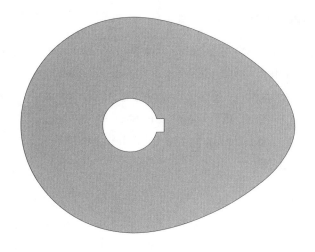

Figure 3:23 Cams

Figure 3:24 Gears & Internal Splines

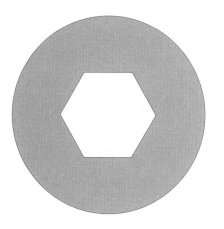

Figure 3:25 Hexes

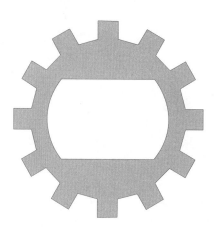

Figure 3:26 Special Shapes

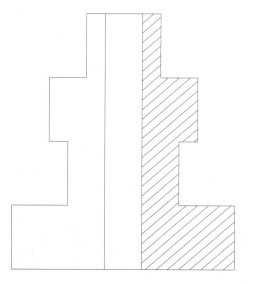

Figure 3:27 Sectionalizing Parts

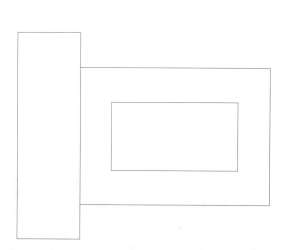

Figure 3:28 Various Shapes in Bars

Cutting Shim Stock Absolutely Burr Free

For most sheet metal parts, lasers are more cost efficient. However, when thin materials need to be cut without many holes, wire EDM can be significantly cheaper and produce an edge that is totally burr free. For example: 500 pieces to be machined from .005 shim stock. With wire EDM, the shim stock is cut and sandwiched between two 1/4 inch (6.4 mm) steel plates, the total height of the shims is 2.5 inches (63.5 mm). See Figure 3:29.

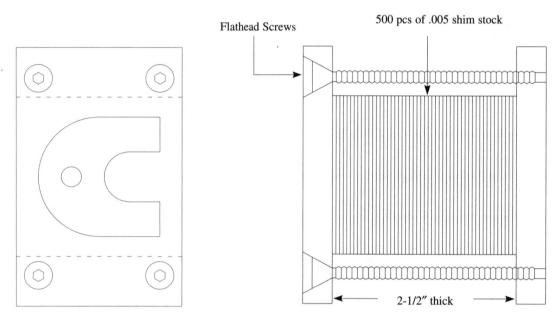

Figure 3:29

EDMing Shim Stock Burr Free

Multiple parts can be cut as shown in Figure 3:30.

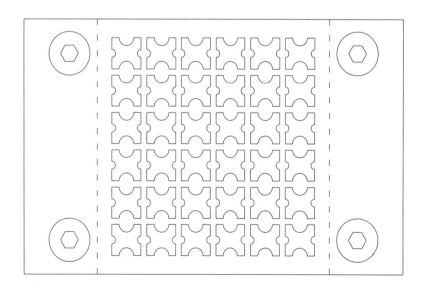

Figure 3:30

Cutting Multiple Shims

Single Cavity Cut Into One Side of a Tube

A large oil field production company needed two tapered cavities to be cut out of one side of a tube. We built a special fixture that enabled us to cut 14" (356mm) deep inside of a tube. Illustrated in Figure 3:31, is a show piece that we cut with the fixture.

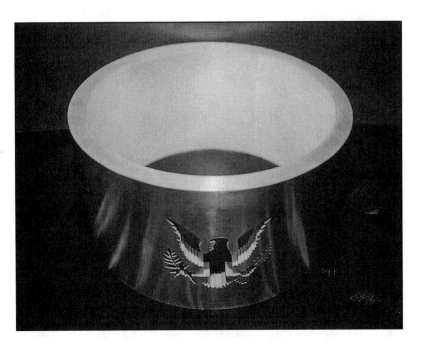

Figure 3:31

Single Cavity Cut Into One Side of a Tube

Materials That Can be EDMed

Wire EDM can cut any electrically-conductive hard or soft material. Some of the materials that can be wire EDMed are listed below.

Inconel	**Aluminum**	**Vasconal 300**
Tool Steels: 01,A2,D2,S7	**Aluminum Bronze**	**PCD Diamond**
Carbide	**Copper**	**Nitronic**
Ferro-Tic	**Brass**	**Beryllium Copper**
CPM 10V	**Cold Roll Steel**	**Hastalloy**
4130	**Hot Rolled Steel**	**Stellite**
Graphite	**Stainless Steels**	**Titanium**

Machining Costs

Usually the machining costs are determined by the amount of square inches of cutting, as illustrated in Figure 3:31.   Other factors are type of material, programming, set up time, and whether the flushing nozzles contact the part.

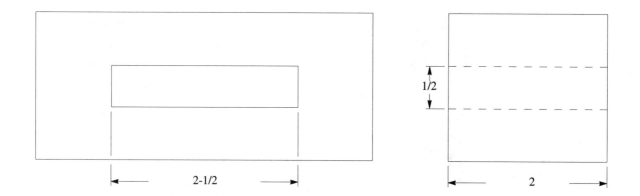

Figure 3:31

Determining Machining Costs
Thickness x Linear Inches = Square Inches
2 x (1/2 + 1/2 + 2 1/2 + 2 1/2) = 12 square inches

This chapter has discussed various profitable uses of EDM. The next chapter will examine the proper procedures for this process.

4 Proper Procedures

To gain the greatest benefits from wire EDM, specific procedures should be used to maximize EDM's potential for reducing machining costs. In planning work, the wire EDM machine can be visualized as a super precision band saw which can cut any hard or soft electrical conductive material.

Starting Methods for Edges and Holes

Three Methods to Pick Up Dimensions.

If the outside edges are important, then a finished edge should be indicated when setting up the part to be wire EDMed.

A. Pick Up Two Edges as in Figure 4:1.

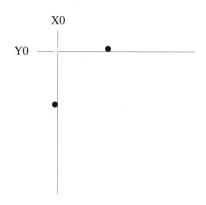

Figure 4:1
Pick Up Two Edges

B. Pick Up a Hole as in Figure 4:2.

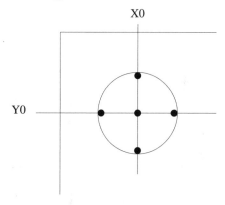

Figure 4:2
Pick Up a Hole

C. Pick Up an Edge and Holes or Two Holes as in Figure 4:3

By using an edge and two holes, a part can be EDMed which is much larger than the capacity of the machine. The part is indicated and a hole that has been either machined or EDMed is picked up. Also, two EDMed edges can be used to locate the part after it has been machined.

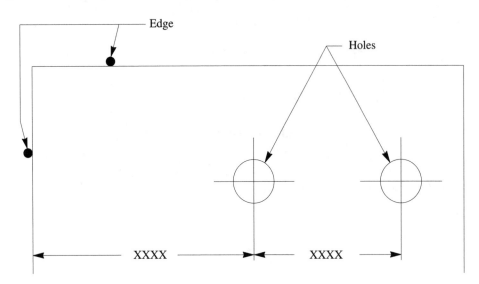

Figure 4:3
Pick-Up From Edges and Holes

Edge Preparation

A. Square Edges

1. Machined or Ground.

To ensure accuracy, the pick up edges must be square as shown in Figure 4:4.

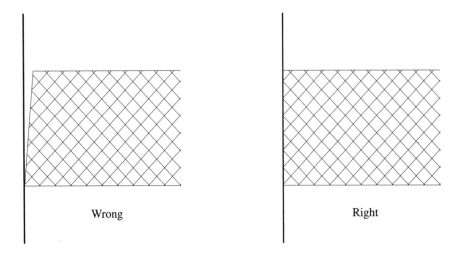

Figure 4:4
Edges must be square for proper pick up.

2. Unfinished Edges.

Parts can be made square to the top surface with sides unfinished, as illustrated in Figure 4:5. The wire is made square to the top surface by means of a special squaring block.

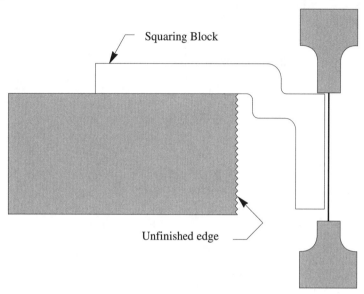

Figure 4:5
Special squaring block can be used to make the wire square
to the surface of the material to be cut.

B. Scale

Since wire EDM is an electrical process, any material that is non-conductive must be removed if it is to be EDMed, or if the area is to be used for picking up. Scale from heat treating is non-conductive. See Figure 4:6.

The heat-treated parts, particularly holes, must be either cleaned of scale or have been vacuum heat treated or wrapped before heat treating. Sand or glass blasting can be used to clean the surfaces where the wire will cut in. However, deep holes are difficult to clean with sand or glass blasting.

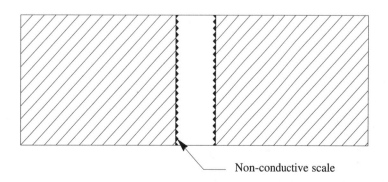

Figure 4:6
To pick up from holes, the holes must be free of scale.

C. Pick-Ups

It is preferred to pick up surfaces without obstructions. If obstructions occur, pick-ups can sometimes be made from a step by means of a gauge block or gauge pin. See Figures 4:7 and 4:8.

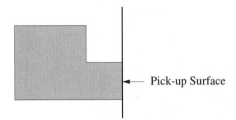

Pick-up Surface

Figure 4:7
Non-obstructive Pick-Up

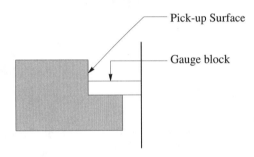

Pick-up Surface

Gauge block

Figure 4:8
Obstruction Pick-Up—A gauge block is used for pick-up.

Starter Holes

A. Automatic Pick-up

When locating parts with starter holes, the machine will automatically pick up the center of the hole as shown in Figure 4:9. Such holes should be free from burrs or scale.

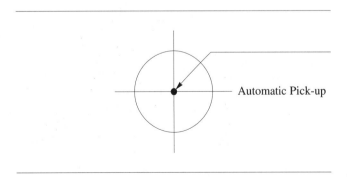

Automatic Pick-up

Figure 4:9
Wire EDM machines automatically pick up the center of a hole.

B. Unsquare Holes

If a hole is unsquare, as illustrated in Figure 4:10, the wire will pick up the high points and not the center of the hole.

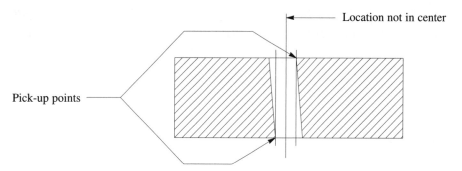

Figure 4:10
Unsquare hole will produce an inaccurate pick-up.

C. Relieved Holes

A relieved hole, as pictured in Figure 4:11 , is the most accurate method to pick up from a hole. Approximately 1/8″ (3 mm) to 1/4″ (6 mm) of land should be left.

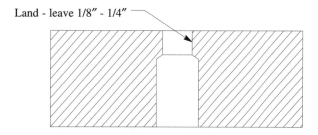

Figure 4:11
The greatest accuracy is obtained with a relieved hole.

D. Smooth Holes

A drilled hole may leave ragged edges. The wire will pick up the high points of the ragged edges. To ensure accuracy, a reamed or bored hole is best. See Figure 4:12 .

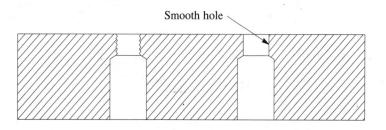

Figure 4:12
Smooth holes locate pick-ups most accurately.

E. Placement and Location of Starter Holes

1. If the part pick-up is in another location, the starter hole requires no precise location.

2. If the starter hole is used for pick-up, then the starter hole should be placed at a straight surface whenever possible, as shown in Figure 4:13. When parts are not skim cut, usually a slightly raised area appears where the part ends. In such cases the tip can be removed with a file or stone.

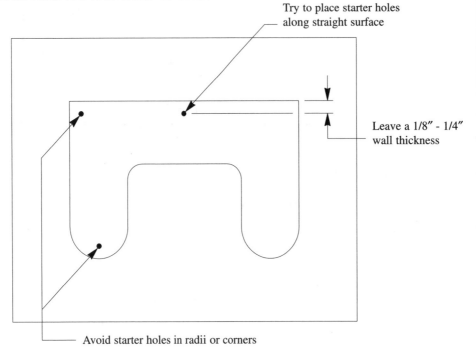

Figure 4:13
Proper Placement of Starter Holes

3. On narrow slots, the starter hole should be placed in a corner, as illustrated in Figure 4:14, so that only one slug will be produced when wire EDMed.

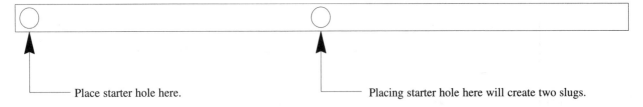

Figure 4:14
Proper Placement of Starter Hole for Narrow Slots

5 Understanding the Wire EDM Process

Accuracy and Tolerances

Wire EDM is extremely accurate. Many machines move in increments of 40 millionths of an inch (.00004") (.001 mm), some in 10 millionths of an inch (.00001") (.00025 mm), and others even in 4 millionths of an inch (.000004") (.0001 mm).

Machines can achieve accuracies of +/-.0001" (.0025 mm). Skim cuts are made to obtain such tolerances. See Figure 5:1.

Courtesy Mitsubishi

Figure 5:1
Precision Wire EDMing

Finishes

Extremely fine finishes of below 15 RMS can be produced with wire EDM. (Some machines can produce even a mirror finish.) Wire EDM produces an excellent finish even in the so-called "rough cut." Customers are often amazed when shown the fine finish of a single-pass cut.

This fine finish is present even after very large parts are cut, as in Figure 5:2. In other cutting operations, such as lasers and abrasive water jet, the larger the part, the rougher the finish. Wire EDM produces a smooth finish because the wire electrode goes through the entire part, and spark erosion occurs along the entire wire electrode.

Figure 5:2
Part 16 Inches High Cut at Reliable EDM (They can cut up to 30 in. tall).

Wire Path

A. Wire Kerf

The wire never contacts the workpiece. The wire electrode cuts by means of spark erosion, leaving a path slightly larger than the wire. A commonly used wire, .012″ (.30 mm), usually creates a .015″ (.38 mm) to .016″ (.41 mm) kerf as shown in Figure 5:3. Thinner wires have smaller kerfs

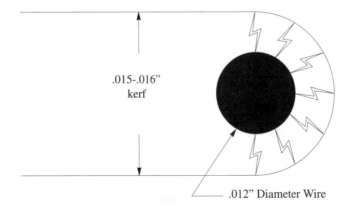

.015-.016″ kerf

.012″ Diameter Wire

Figure 5:3
Wire Kerf

B. Corners and Radii

When the wire turns a corner, it can produce a sharp edge on the outside corner, but it will always leave a small radius on the inside corner as demonstrated in Figure 5:4. The size of this radius is determined by the wire diameter plus the spark gap.

To produce very sharp outside corners, skim cuts are made. Having small corner radii on the outside corners can prevent the need for skim cuts; this also reduces wire EDM costs. In stamping dies, sharp corners usually wear first, so a small outside radius is preferable.

The minimum inside radius for .012″ (.30 mm) wire is .007″ (.018 mm), and the minimum radius for .010″ wire (.25 mm) is .006″ (.15 mm). Smaller radii are possible with thinner wire; however, most work is done with thicker wires because thinner wire cuts slower.

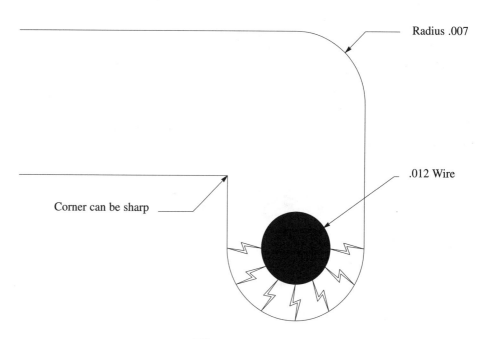

Radius .007

.012 Wire

Corner can be sharp

Figure 5:4
Inside and Outside Corners

Skim Cutting

For most jobs, the initial cut is sufficient for both finish and accuracy. However, for precision parts, skim cuts achieve greater accuracy and a finer finish. There are three main reasons for skim cuts:
- barreling effect and wire trail-off
- metal movement
- finishes and accuracy

A. Barreling Effect and Wire Trail-Off

There is a .001″ to .002″ (.025-.05 mm) gap between the wire and the workpiece. In this gap, a controlled localized eruption takes place. The force of the spark and the gases trying to escape causes a slight barreling. On thick workpieces, this barreling causes the center to be slightly hollow. See Figure 5:5.

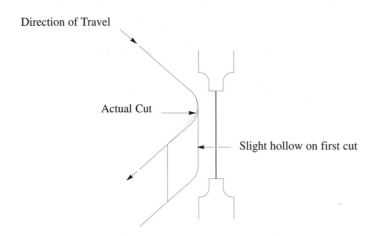

Direction of Travel

Actual Cut

Slight hollow on first cut

Figure 5:5
First Cut Corner Conditions

When cutting sharp corners, the wire dwells longer by the inside radius, causing a slight overcut; on the outside radius, it speeds, leaving a slight undercut as illustrated in Figure 5:6.

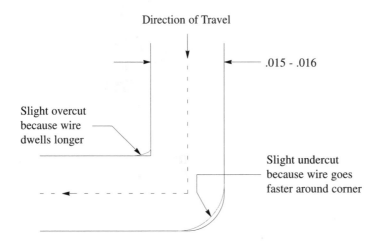

Direction of Travel

.015 - .016

Slight overcut because wire dwells longer

Slight undercut because wire goes faster around corner

Figure 5:6
Skim Cutting is Used For Very Close Tolerances.

A trail-off is produced when the machine cuts a corner. A slight amount of material is left behind for a short distance before the wire returns to its programmed path. For most jobs, this slight undercut is negligible.

The sharper the corner, the greater the overcut and undercut. The accuracy of the part determines the need for skim cutting.

To avoid most of this barreling effect and wire trail-off, some wire EDM machines automatically slow down in corner cutting. Nevertheless, high precision parts still require skim cuts.

B. Metal Movement

Even though metal has been stress relieved, it may move after the part has been cut with wire EDM because the stresses within the metal were not totally removed in stress relieving. If metal has moved due to inherent stresses, and the part requires to be precise, then skim cuts are needed to bring the part into tolerance. The accuracies called for by the print determine the number of skim cuts.

C. Finishes and Accuracy

First cuts produce a fine finish; however, sometimes a finer finish and greater accuracies are required. To accomplish this, skim cuts are used. See Figure 5:7 for a general view of the various finishes that can be produced with wire EDM. (Some machines produce different results.)

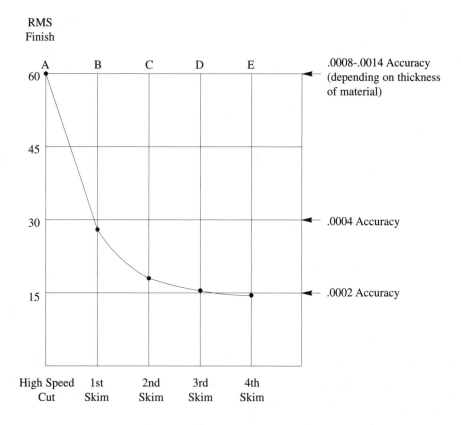

Figure 5:7
Cut A—For most jobs, this finish and accuracy are more than adequate.
Cuts B-E—Depending on accuracy and finish
required, various skim cuts are performed.

Skim cutting produces fine finishes because less energy is applied to the wire, thereby creating smaller sparks and thus smaller cavities. These small sparks produce extremely fine finishes, and on some machines a mirror finish.

Carbide

Tungsten carbide, third in hardness to diamond and boron carbide, is an extremely difficult material to machine. Except for diamond cutting tools and diamond impregnated grinding wheels, EDM presents the only practical method to machine this hardened material.

To bind tungsten carbide when it is sintered, cobalt is added. The amount of cobalt, from 6 to 15 percent, determines the hardness and toughness of the carbide. The electrical conductivity of cobalt exceeds that of tungsten, so EDM erodes the cobalt binder in tungsten carbide. The carbide granules fall out of the compound during cutting, so the amount of cobalt binder determines the wire EDM speed, and the energy applied during the cutting determines the depth of binder that is removed.

When cutting carbide on certain wire EDM machines, the initial first cut can cause surface micro-cracks. To eliminate them, skim cuts are used. However, at our company, Reliable EDM, we have repeatedly cut carbide parts with a single cut. When precision carbide parts are needed, then skim cuts are used.

Some older wire EDM machines used capacitors. Since these machines applied more energy into the cut, there was a greater danger for surface micro-cracking. Then DC power supply machines without capacitors were introduced, and this helped in producing less surface damage when cutting carbide.

Today, many machines come equipped with AC power supplies. These machines are especially beneficial when cutting carbide in that they produce smaller heat-affected zones and cause less cobalt depletion than DC power-supplied machines.

To eliminate any danger from micro-cracking and to produce the best surface edge for stamping, it is a good practice to use sufficient skim cuts when EDMing high-precision blanking carbide dies. Studies show that careful skimming greatly improves carbide surface quality. Durability tests prove that an initial fast cut and fast skimming cuts produce very accurate high-performance dies.

Polycrystalline Diamond

The introduction of polycrystalline diamond (PCD) on a tungsten carbide substrate has greatly increased cutting efficiency. PCD is a man-made diamond crystal that is sintered with cobalt at very high temperatures and under great pressure. The tungsten substrate provides support for the thin diamond layer.

The cobalt in PCD does not act as a binder, but rather as a catalyst for the diamond crystals. In addition, the electrical conductivity of the cobalt allows PCD to be EDMed. When PCD is EDMed, only the cobalt between the diamonds crystals is being EDMed.

EDMing PCD, like EDMing carbide, is much slower than cutting steel. Cutting speed for PCD depends upon the amount of cobalt that has been sintered with the diamond crystals and the particle size of PCD. Large particles of PCD require very high open voltage for it to be cut. Also, some power supplies cut PCD better than others.

Ceramics

Ceramics are poor conductors of electricity. However, certain ceramics are formulated to be cut with wire EDM.

Flushing

Flushing is an important factor in cutting efficiently with wire EDM. Flushing pressure is produced from both the top and bottom flushing nozzles. See Figure 5:8. The pressurized deionized fluid aids in spark production and in eroded metal-particle removal.

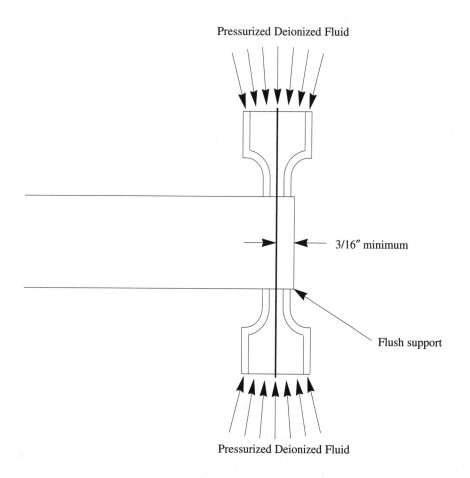

Figure 5:8
Ideal Flushing Conditions

Sometimes the flushing nozzle may extend beyond the edge of a workpiece, as shown in Figure 5:9. When this occurs, flushing pressure is lost, and this can cause wire breakage and part inaccuracy. To avoid wire breakage in such cases, a lower spark energy is used which slows the machining process. To avoid losing flushing pressure, it is advisable, if possible, to leave at least 3/16″ (5 mm) of material to support the flushing nozzles.

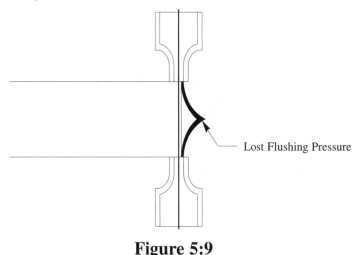

— Lost Flushing Pressure

Figure 5:9
Poor Flushing Conditions

Cutting Speed

Speed is rated by the square inches of material that are cut in one hour. Manufacturers rate their equipment under ideal conditions, usually 2 1/4 inch (57 mm) thick D2 hardened tool steel under perfect flushing conditions. However, differences in thicknesses, materials, and required accuracies can greatly alter the speeds of EDM machines.

Cutting speed varies according to the conductivity and the melting properties of materials. For example, aluminum, a good conductor with a low melting temperature, cuts much faster than steel.

On the other hand, carbide, a non-conductor, cuts much slower than steel. It is the binder, usually cobalt, that is melted away. When the cobalt is eroded, it causes the carbides to fall out. Various carbides machine at different speeds because of carbide grain size and the binder amount and type.

Impurities

Generally, impurities cause little difficulty; however, occasionally materials are received with non-conductive impurities. The wire electrode will either stall or pass around small non-conductive impurities, thereby causing possible streaks from raised or indented surfaces.

When welded parts must be EDMed, use caution to make certain there is no slag within the weld. Tig welding is preferred for wire EDM.

Recast and Heat-Affected Zones

The EDM process uses heat from electrical sparks to cut the material. The sparks create a heat-affected zone that contains a thin layer of recast, also called "white layer". The depth of the heat-affected zone and recast depends upon the power, type of power supply, and the number of skim cuts.

The recast contains a layer of unexpelled molten material. When skim cuts are used, much less energy is applied to the surface, which greatly reduces and practically eliminates the recast layer.

On older wire EDM machines, the heat-affected zones and recast were much more of a problem. Also, the recast and heat-affected zones of ram EDM are much greater when roughing because more energy can be used than with wire EDM.

Many of today's wire EDM machines have reduced this problem of recast and heat-affected zones. Our company, Reliable EDM, is a wire EDM job shop that has done work for well over 500 companies, including aerospace companies. We have wire EDMed thousands of jobs and cut all sorts of materials, including carbide and high-alloy steels. We have had practically no negative results from recast and heat-affected zones. Most work is done with just one cut. For precision parts, skim cuts are used.

Our newer machines now come equipped with anti-electrolysis power supplies, also called AC power supplies. These power supplies greatly reduce the recast and heat-affected zones. On some machines, the heat-affected zone for the first cut is .0015″ (038 mm), on the first skim cut it is .0003″ (.0076 mm), and on the second skim cut it is .0001″ (.0025 mm).

For years, the recast and heat-affected zones have been a concern for the aerospace and aircraft industry. With the improvement of power supplies, these industries increasingly accept work done with wire EDM.

AC Non-Electrolysis Power Supplies

Instead of cutting with DC (direct current), some machines cut with AC (alternating current). Cutting with AC allows more heat to be absorbed by the wire instead of the workpiece.

Since AC constantly reverses the polarity of the electrical current, it reduces the heat-affected zone and eliminates electrolysis. Electrolysis is the stray electrical current that occurs when cutting with wire EDM. For most purposes, electrolysis does not have any significant affect on the material. However, the elimination of electrolysis is particularly beneficial when cutting precision carbide dies in that it reduces cobalt depletion.

When titanium is cut with a DC power supply, there is a blue color along where the material was cut. This blue line is not caused by heat, as some suspect, but by electrolysis. This effect is not generally detrimental to the material. However, AC power supplies eliminates this line.

Like AC power supply, the AE (anti-electrolysis) or EF (electrolysis-free) power supplies improve the surface finish of parts by reducing rust and oxidizing effects of wire EDM. Also, less cobalt binder depletion occurs when cutting carbide, and it eliminates the production of blue lines when cutting titanium. AC and non-electrolysis power supplies definitely have advantages.

Heat-Treated Steels

Wire EDM will machine hard or soft steel; however, steel in the hardened condition cuts slightly faster. Materials requiring hardening are commonly heat treated before being cut with wire. By heat treating steel beforehand, it eliminates the distortions that can be created from heat-treating.

Cutting Large Sections

Steels from mills have inherent stresses. Even hardened steel that has been tempered often has stresses remaining. For cutting small sections, the effect is negligible. However, for large sections when there is a danger of metal movement, it is advisable to remove some of the metal. By removing metal, it reduces the possibility of metal movement. See Figure 5:10.

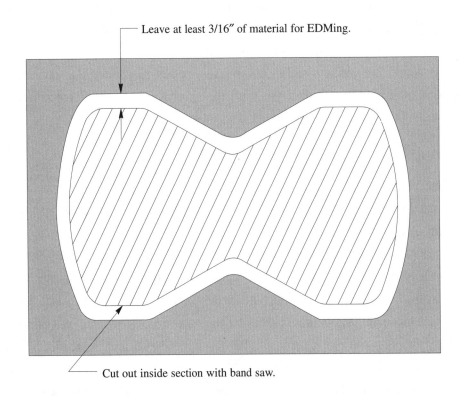

Leave at least 3/16″ of material for EDMing.

Cut out inside section with band saw.

Figure 5:10
Removing Material to Reduce Stresses on Large Parts

Metal movement can also be reduced by cutting relieving slots with a band saw to connecting holes, as in Figure 5:11. Steel should be stress relieved before heat-treating to remove the stresses caused by milling, drilling, and grinding. After heat-treating, the tool steel should be double or triple drawn, including the non-deforming air hardening tool steels. Another method to remove stresses is to use cryogenics (deep freeze). The tool steel is hardened and tempered; then it is put into deep freeze and retempered. It is also best that the top and bottom surfaces of the plate are ground.

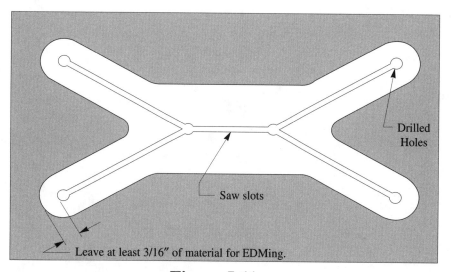

Figure 5:11
Using Saw Slots to Reduce Stresses on Large Parts

Cutting Sections From a Block

A. Leaving a Frame

When a section must be cut from a block of steel, a frame should be left around the workpiece to ensure accuracy and to reduce cost. At least 1/4 to 1/2″ (6.5 to 13 mm) should be left around the part so that flush nozzles can efficiently remove the eroded particles and also support the part for clamping. See Figure 5:12.

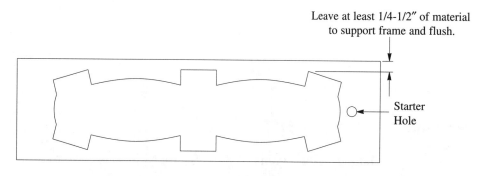

Figure 5:12
Support Part With Frame

B. Strength of Frame

Sufficient extra material needs to be left around the part. When the part is held in a fixture, the extra material will prevent the part from moving as it is being EDMed. Figure 5:13 demonstrates a weak support frame. While the part is being EDMed, the frame becomes weak, which can cause the part to move.

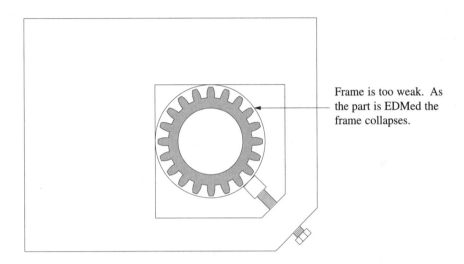

Frame is too weak. As the part is EDMed the frame collapses.

Figure 5:13
Improper Support Frame

C. Material for Clamping

For many parts fixtures are used as in the above illustration. However, for some parts provision should be made for clamping. See Figure 5:14.

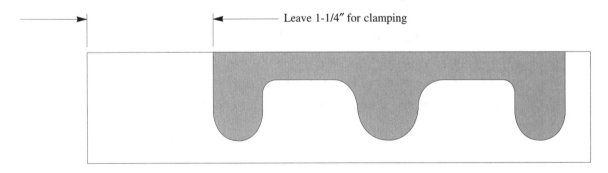

Leave 1-1/4″ for clamping

Figure 5:14
Extra material provided for clamping

Understanding the Wire EDM Process

The better understanding one gains of the wire EDM process, the more benefits one can obtain from this amazing process. The next section covers how to reduce wire EDM costs.

6 Reducing Wire EDM Costs

Wire EDM costs can be greatly reduced if the material has been properly prepared and the EDM process is understood. Unfortunately, the opposite is also true. Wrong preparation can be costly.

Create One Slug

To reduce costs, the general aim should be to create one slug. Wire EDM is an automatic process; if more slugs are made, it requires more down time and operator services. Also, when surfaces close to an edge are cut, inadequate flushing occurs which reduces cutting speed.

When entering a workpiece on non-submerged machines, lower power settings and slower cutting speeds must be used because of poor flushing conditions. Also when entering a part at a slight angle, feathered-edge machining occurs. This feather-edge machining may cause slight surface irregularities. Skim cutting can be used to remove such irregularities; however, unnecessary skim cuts increase cost. Cutting one slug is much more cost effective. See Figures 6:1 - 6:4.

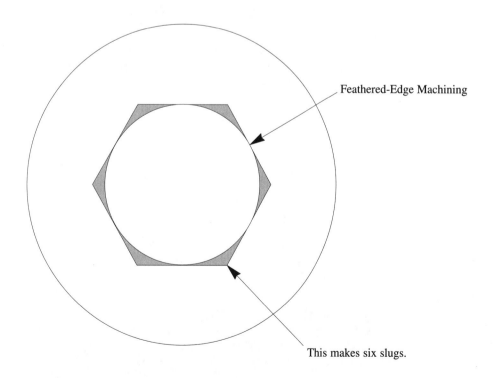

Feathered-Edge Machining

This makes six slugs.

Figure 6:1
Wrong Procedure—Creates six slugs, and slows the process with feather-edge machining

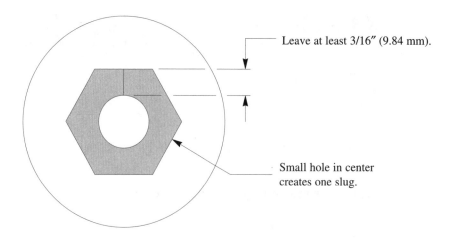

Leave at least 3/16″ (9.84 mm).

Small hole in center
creates one slug.

Figure 6:2
Right Procedure—Creates one slug which produces more efficient machining

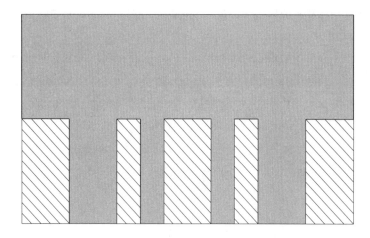

Figure 6:3
**Wrong Procedure—Creates Five Slugs—Five starting cuts must be made,
and five times the machine must be stopped to remove each slug.**

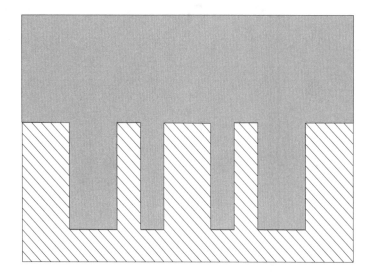

Figure 6:4
**Right Procedure—Creates One Slug—Leaving extra material
on the outside allows for one slug to be cut.**

Keeping Flush Ports on the Workpiece

The most efficient method for wire EDMing is placing both top and bottom flush ports on the workpiece as shown in Figure 6:5. This placement allows for maximum flushing pressure to remove the eroded chips.

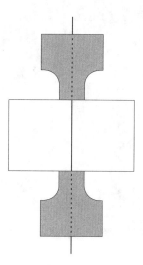

Figure 6:5
Most efficient cutting occurs when both flush ports rest on the part.

If possible, air machining should be avoided. It is less efficient because the flush ports are not resting on the workpiece (See Figure 6:6).

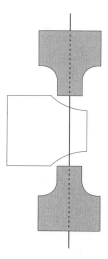

Figure 6:6
Air Machining is Possible But Less Efficient

For many applications, however, there is no alternative to air machining. At our company, Reliable EDM, we cut many jobs in the air, including tall parts, because there is no other alternative. See Figure 6:7.

Figure 6:7
An 11-1/2 Inch (292 mm) test specimen cut in the air out of a large gear

Machining After Wire EDM

To avoid air machining, it is sometimes more economical to do the machining after, rather than before the EDM process. This is particularly true with shallow recesses as in Figure 6:8.

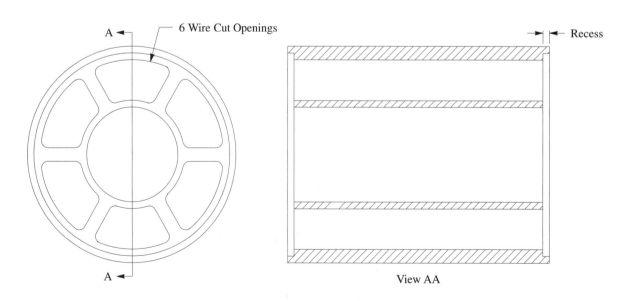

Figure 6:8
Machine the Workpiece After Wire EDMing
Since the recess is shallow, it is more efficient to do the EDMing when the part is solid.

Often parts are stacked to reduce costs. When parts have intricate dimensions, stacking may be difficult if parts have been previously machined as shown in Figure 6:9.

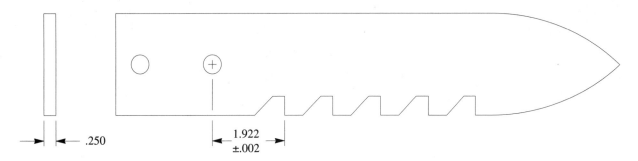

.250   1.922
                        ±.002

Figure 6:9
Holes should be put in after EDMing.
Making one piece presents no problem; however, parts like these are stacked. If holes are premachined, it is difficult to line up the holes when cutting large stacks.

If parts can be stacked, it is preferred that holes be put in after the part has been EDMed. Putting holes in first can cause alignment difficulties when the parts are set up in a fixture. See Figure 6:10.

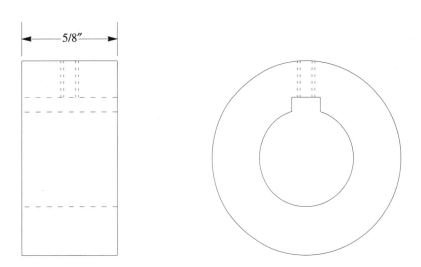

5/8″

Figure 6:10
Put tapped hole in after EDMing.
Parts like these are often stacked in a "V" block. Higher machining costs occur because tapped holes cause alignment difficulties.

Cutting Multiple Plates and Sheet Metal Parts

Stacked sheet metal can be held with fixtures without the need for welding. However, when multiple parts from one stack and starter holes are required, the stack can be bolted with flat head screws or welded on its sides. The stack should be flat, and the EDM job shop should be consulted for the ideal stack thickness.

Accuracy, efficiency, and machine capabilities determine the height for stacked parts.

Wire EDM will cut through light rust; however, heavy rust and scale must be removed.

Many times plates are warped. The plates should be clamped tightly together before welding. At least 1/2″ (13 mm) should be left on the sides for welding and clamping the part. See Figure 6:11 for proper stacking.

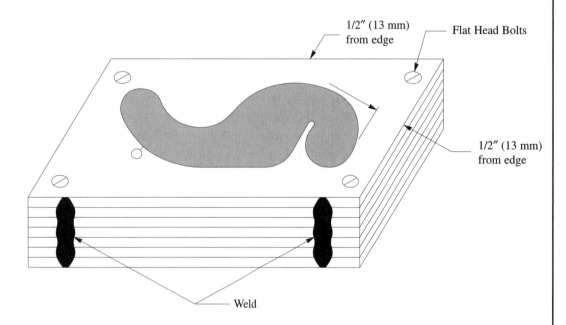

Figure 6:11
Stacks Welded or Bolted
At least 1/2″ (13 mm) should be left for clamping and a frame to support part while cutting.
Caution: If parts are welded or bolted, both sides of plates must be clean and
free from heavy scale, tape, paper, or any other non-conductive materials.

If sheets or plates are badly warped, each stack should be divided in half and the belly should hit the center. The ends are then clamped together and welded. The aim should be to produce a flat surface. The weld should be removed from the top and bottom of the stack so flush ports do not hit the weld.

When putting stacks together, all sheets must be clean—marker paint, (not magic marker), scale, tape, or paper between the sheets must be removed. Wire EDM cuts by spark erosion; it cannot cut through non-conductive materials.

Production Lots

Wire EDM is an excellent machining method for production work. Fixtures are often used to hold the multiple parts. It is important that production lots are machined the same in the area where they will be located. Parts also need to be machined square. See Figure 6:12.

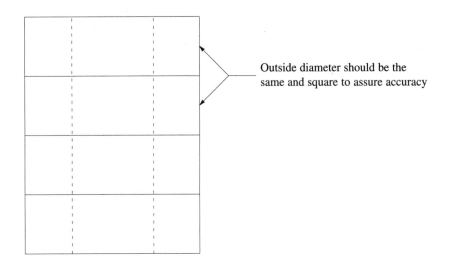

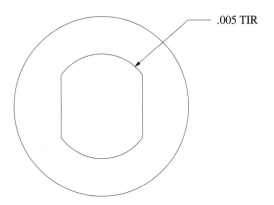

Figure 6:12
Production EDMing—When machining parts, consideration should be made for stacking.

Stipulating Wire Sizes

Some machines can cut with .0012″ (.030 mm) wire. One wire EDM job was done on a .015″ (.38 mm) diameter air turbine rotor. It had 13 slots cut with .00039″ (.01 mm) wire. This was done on a specialized wire EDM machine. For some companies, .004″ (.1 mm), a little less than two hair thicknesses, is the limit of the wire they use for cutting.

The difficulty with cutting with thin wires is that it machines much slower because less energy can be applied to the wire. Also, thin wires break much more easily than standard wire sizes.

Some applications require thin wires; however, whenever possible, stay with the standard wire size of .010″ (.25 mm) or .012″ (30 mm) wires. Stipulating thin wire can add significant costs to the wire EDM process because of slower cutting feeds and difficulties associated with such wires.

Pre-Machining Non-Complicated Shapes

It is not always necessary to EDM the entire part. Sometimes pre-machining can reduce costs as shown in Figure 6:13.

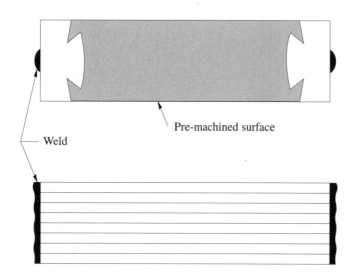

Figure 6:13
Pre-Machine Parts to Reduce Costs.

Wire EDM is an extremely efficient method to machine parts. However, costs can be further reduced by understanding this process.

The introduction of wire EDM into manufacturing has greatly altered the machining process as will be shown in the next chapter.

7 Wire EDM Applications

Production Wire EDM

Before starting Reliable EDM, I worked as a tool and die maker and as an operations manager of a large tool and die stamping company. We had wire EDM, and I saw its great capabilities for making individual parts. However, when I began Reliable EDM, I did not realize that the increased cutting speeds of wire EDM made us often more cost effective in machining production parts than companies with their CNC equipment.

These companies would send their prints, and we would quote the parts to be EDMed. We began to get more and more production EDM. Today our dominant work is production wire EDM. See Figures 7:1-7:5.

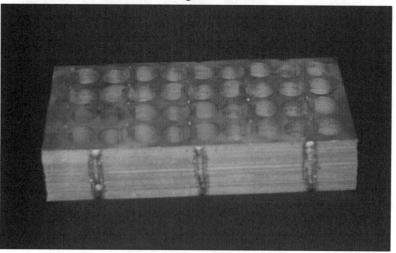

Figure 7:1
Precision Cams Cut From Stainless Steel Sheets

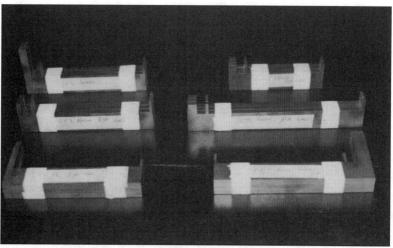

Figure 7:2
Short-Run Production of Multiple Gages

Figure 7:3
Production Cutting of Machine Parts From 17-4 Stainless Steel

Figure 7:4
Special Fixture to Cut Multiple Machine Parts

Figure 7:5
Titanium Parts for Oil Exploration

A Great Problem

A great problem is that many designers and engineers are unaware of the great potential of wire EDM to machine parts. As more and more engineers, tool designers, and machinists discover the capabilities of wire EDM, they discover that wire EDM is a very cost effective and accurate machining process. One of the major tasks Reliable EDM has undertaken is to educate industry in the capabilities and advantages of wire EDM.

Examples of Wire EDM

Understanding the capabilities of wire EDM permits many unique applications. As specialists in wire EDM, our company has done thousands of jobs for hundreds of companies. Following are examples of work done by Reliable EDM:

A. Tall Parts

Figure 7:6
Flange Cutout—Slug is on top of the flange.
Four angular cutouts were made on a large shaft.

B. Modified Machines

At our company we have modified a number of our machines. Illustrated in Figure 7:7 is a machine we modified to cut 19 1/2" (495 mm). To the right of it is a 30" (762 mm) bar of titanium we wire EDMed on another machine. Due to customer confidentiality, we are not showing a 37" (940 mm) cut we did in a bar of beryllium copper. We modified this machine to cut 38" (965 mm).

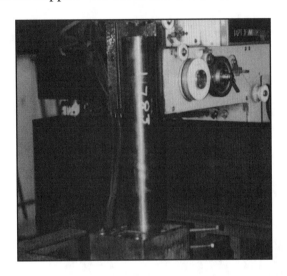

Figure 7:7
Machines Modified: Left--Cut 19-1/2 Inches Tall (495 mm)
Right: Titanium Cut 30 inches tall (762 mm). Ruler: 36" (914mm)

Submersible Cutting

We also did a slight modification on another machine to cut 18" (457 mm) submersed. See Figure 7:8.

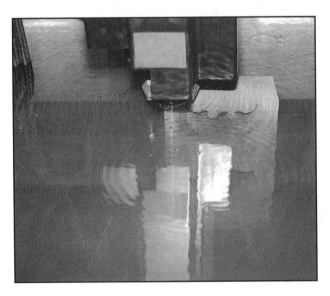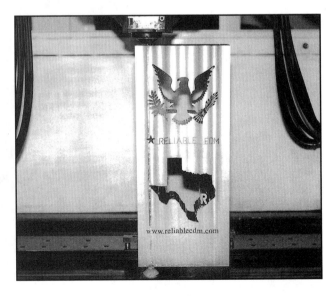

Figure 7:8
Submersible Cutting to 18" (457 mm)

Figure 7:9
EDMing a 24 Inch Mold—To cut this 24 inch mold,
part of the mold was EDMed and then turned.

C. Large Heavy Gears

Figure 7:10
Cutting Two Keyways Into a 54 Inch (1,372 mm) Diameter Gear Weighing 3,000 Pounds

D. Various Tall Parts

Figure 7:11
Various Tall Parts EDMed—In front is a 6 inch vernier caliber.

E. Overhanging Parts

Serrations Cut Into a Large Shear Blade.

Figure 7:12
Cutting a Special Shear Blade

F. Long Tubes

Figure 7:13
Splitting a Long Tube

E. Other Applications

Cutting a Large Gate Valve.

Figure 7:14
Large Gate Valve—Ruler is 24 inches (610mm)

Cutting Cavities in One Side of a Tube.

Figure 7:15
Fixture Made to Cut a Cavity in One Side of a Tube

Large Carbide Rock Bit Being Cut in the Air.

Figure 7:16
Test Specimen Being Cut From a Carbide Rock Bit Measuring Over 20″ (Ruler is 24″)

Cutting an Inside Diameter and a Keyway of a Large Gear.

Figure 7:17
One-Cut: Inside Diameter and Keyway

Splitting Machined Parts

If parts need to be split with traditional machining, the other half is often destroyed. With wire EDM, the part can be spilt leaving a kerf of just the thickness of the wire, plus .002 to .003" (.05 to.076 mm) A .010" (.25 mm) wire leaves a kerf from .012 to .013" (.30 to.33 mm) kerf. Our company has split thousands of jobs.

Figure 7:18
Machined Parts Split in Half

Cutting a Test Specimen

Wire EDM is ideal for cutting test specimens because it creates no stresses in the material and it produces an excellent finish. Figure 7:19 shows our cutting of an 11-1/2 inch (292 mm) test specimen out of this large gear. For the entire cut the flush nozzles never came in contact with the gear.

Figure 7:19
In-the-air cutting to obtain a test specimen (11 1/2" 292 mm) from a large gear

Advantages of Wire EDM for Die Making

A. Old-Fashioned Tool and Die Making

Wire EDM has revolutionized tool and die making. To understand the extent of the wire EDM revolution in tool and die making, let me share some history. In 1950, I started to work in a machine shop; one year later I became an apprentice tool and die maker in a large handbag frame plant in Brooklyn. The plant produced a large variety of handbag frames which required many kinds of fixtures and dies.

From 01 tool steel we milled, ground, or filed the form punch. The punch was then hardened in a gas-fired oven that had no temperature gauge. In those days, one learned early the necessary cherry red color to indicate that the punch was ready to be quenched in oil. After quenching, we used a gas torch to temper the punch to a light straw color.

Using the hardened punch as a template, we traced the pattern on a piece of tool steel colored with Dykem blue. We used a band saw to cut as close to the line as possible. We placed the hardened punch on top of the soft die section and placed both of them under a power press. The power press was bounced by hand until we made an indentation into the soft die section.

Then we used a filing machine and hand files to remove the excess material. We brushed Dykem blue into the cavity, and the punch and die were again placed in the press to make a further indentation. Then the workpiece went back to the filing machine. We repeated this process over and over until a proper fit was made. Then we set the filing machine on an angle to produce the proper taper. The die was then hardened with hopes that the 01 tool steel would not distort when quenched in oil.

Then I took another position in a precision die shop in Long Island City, Queens, N Y. This shop was a new world of die making—here we ground the die sections to exact specifications; some within .0001″ (.0025 mm).

To make these dies we had no comparator or optical equipment. One worker used a large magnifier to check his die work for the proper clearance; but this made his eyes bloodshot from constantly looking through the magnifier. When my turn came to make these dies, I decided to grind the punch and die sections to precise dimensions. Instead of constantly relying on sight, I used a tenth indicator and gauge blocks to obtain the proper dimensions. This is one of the notes I wrote making a precision die: "Grind flat with .016 radius (.40 mm). Move cross feed .001 (.025 mm) at a time. Leave .0003 (.0076 mm) over for finish grinding. Then touch front and back off .0002 (.005 mm), then grind flat. Use dresser three times." See Figure 7:20.

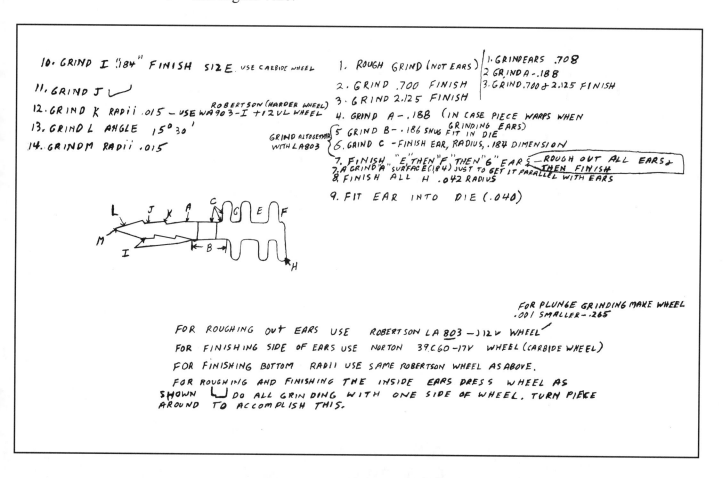

Figure 7:20
Author's handwritten shop sketch for grinding floral pick punches and dies.
These dies ran continuously. The floral picks went into automatic dispensers, so no burrs were tolerated.
The called-for clearance was between .0005″ (.013 mm) to .001″ (.025 mm).

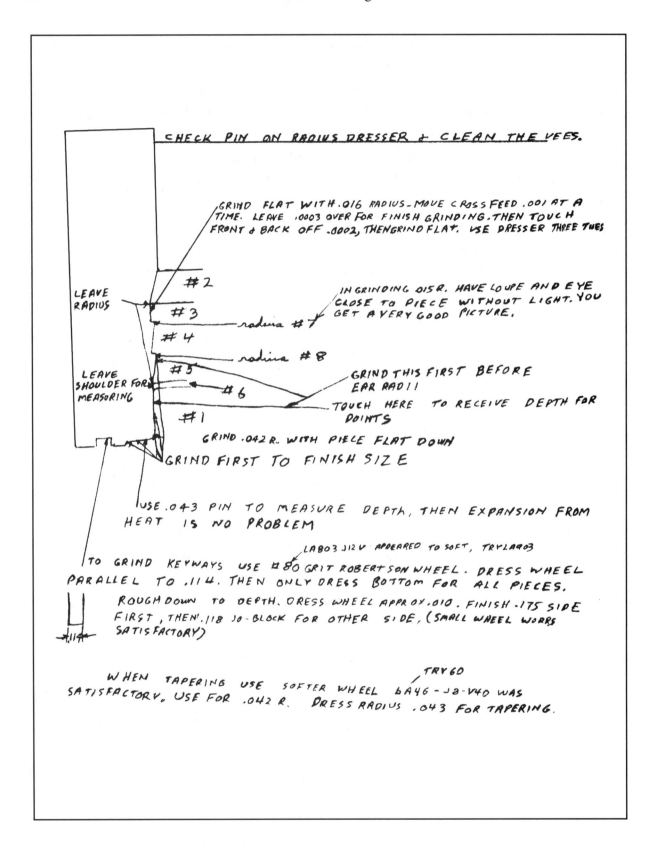

Figure 7:21
Old-Fashioned Precision Die Making
Note the tight tolerances the author wrote for grinding the tip of the floral pick die section: "Grind flat with .016 (.40 mm) radius. Move crossfeed .001 (.025 mm) at a time. Leave .0003 (.0076 mm) over for finish grinding. Then touch front @ back off .0002 (.005 mm), then grind flat. Use dresser three times."

B. The Revolution

To produce these precision dies, it required highly skilled tool and die makers. Then came wire EDM. Now making a computer program of the shape, the production of a much better and more accurate tool was possible.

Tool and die makers are still needed to assemble tooling, but wire EDM has eliminated the need for those skilled die makers to make the many elaborate punch and die sections. Today, wire EDM performs that costly and laborious job. As a result, it has greatly reduced tooling costs, and at the same time produced a superior quality die.

C. Advantages of Wire EDM Dies

1. One-Piece Die Sections

Previously complicated dies were sectionalized—this allowed the die sections to move. See Figure 7:22. Now with wire EDM, the die can be made from a solid block of tool steel producing a much more rigid die, as in Figure 7:23. In addition, sectionalized dies require much more mounting time than a one-piece die section.

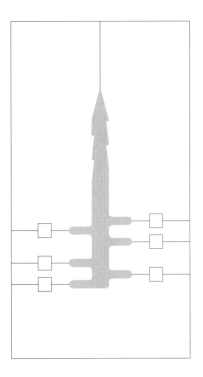

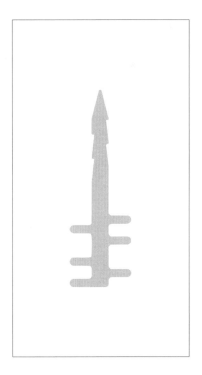

Figure 7:22
Sectionalized Die Sections

Figure 7:23
Solid Die Section

**Wire EDM eliminates costly sectionalized dies and
produces a superior and less costly solid dies.**

2. Exact Spare Parts

To keep up production, spare sections can be on hand in case of wear or breakage. Since computer programs can be stored, spare sections can be precisely duplicated without having the previous part.

3. Dowel Holes EDMed

When die sections or punches need to be changed due to wear or design change, dowel holes can also be EDMed. This produces exactly duplicated replacement die or punch sections.

4. Better Tool Steels

With wire EDM, dies and punches can be made with tougher tool steels, even tungsten carbide. These tougher tool steels produce much longer tool life.

5. Accuracy

Many wire EDM machines move in increments of at least 40 millionths of an inch (.00004"—.001 mm); therefore they can maintain accurate forms and clearances.

6. Die Repairs

Broken dies can be saved by replacing the damaged section with a wire EDMed insert, or the damaged area can be hard welded and then wire EDMed. See Figures 7:24 and 7:25

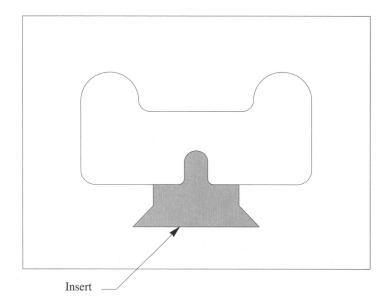

Insert

Figure 7:24
Damaged Die Section Repaired With an Insert

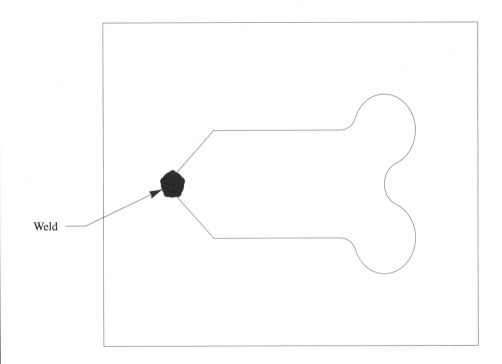

Figure 7:25
Damaged Die Section Repaired With Welding and EDMing

7. Fine Textured Finish

The fine textured surface produced from wire EDM produces longer tool life because of improved surface retention of lubricant.

8. Eliminates Distortion

Punch and dies can be wire EDMed after heat-treatment.. This eliminates the distortions that are created in heat-treating.

9. Inserts for High Wear Areas

If certain areas in the die have a larger wear ratio, inserts can be designed for these wear areas. Then instead of sharpening the entire die, inserts can be installed even with the die in the press.

10. Smaller Dies

Wire EDM allows the building of smaller progressive dies, thereby reducing costs.

11. Longer Lasting

A die lasts only as long as its weakest link. Dies last longer because wire EDM produces exact die clearance which allows the dies to last longer between sharpening, and to be sharpened much farther into the die sections.

12. Punches and Dies From One Piece of Tool Steel

A punch and die can be produced from one piece of tool steel. See Figure 7:26. (The author and his son have written the book, *Wire EDM Handbook.* * This book provides detailed instructions for building one-piece dies, including compound dies from one piece of tool steel. The handbook also describes the many ways to hold stick punches that have been EDMed.)

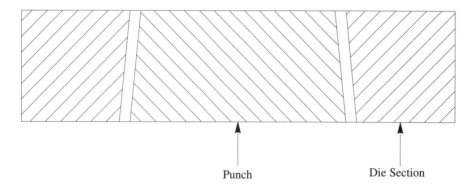

Punch   Die Section

Figure 7:26
Punch and Die Made From One Piece of Tool Steel

13. Cutting Stripper and Die Section Together

Often the stripper may be mounted on the bottom of the die section and cut simultaneously with the die section as shown in Figure 7:27. This significantly reduces the cost when strippers are required.

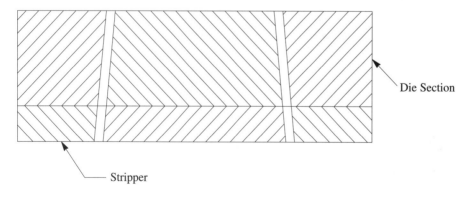

Die Section

Stripper

Figure 7:27
Cutting the Stripper and Die Section Together

(**Wire EDM Handbook*, List price $34.95 plus $3.95 for shipping and handling. This book can be purchased from various sources or from Advance Publishing, Phone (713) 692-0600. Note: Much of the material in this book is a duplication of these chapters on wire EDM. However, the handbook includes extensive information for EDMing stick punches and for building high-performance one-piece stamping dies.)

Wave of the Future

Wire EDM has revolutionized machining. With today's high-speed cutting machines, wire EDM will increasingly replace work performed with traditional methods.

Today, manufacturers, designers, engineers, and those responsible for determining machining methods should endeavor to understand the wire EDM process in order to maximize its great potential. Their knowledge of this process will result in their company saving money, time, and effort while increasing quality product.

Unit 3

Ram EDM

Notes

8 Fundamentals of Ram EDM

Ram EDM Machining

Ram electrical discharge machining (EDM), also known as conventional EDM, sinker EDM, die sinker, vertical EDM, and plunge EDM, as shown in Figure 8:1, is generally used to produce blind cavities.

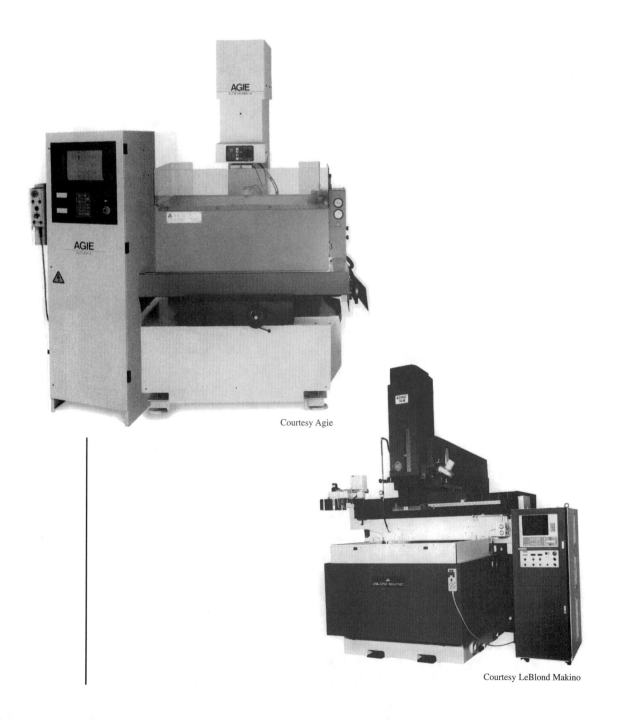

Courtesy Agie

Courtesy LeBlond Makino

Courtesy Charmilles Technologies

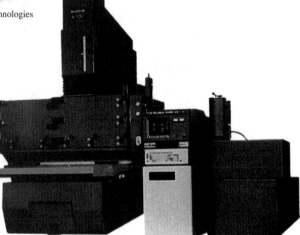

Courtesy Sodick

Courtesy Mitsubishi

Figure 8:1
Ram Electrical Discharge Machines

When a blind cavity is required, a formed electrode is machined to the desired shape. Then by means of electrical current the formed electrode, surrounded by dielectric oil, reproduces its shape in the workpiece, as illustrated in Figure 8:2.

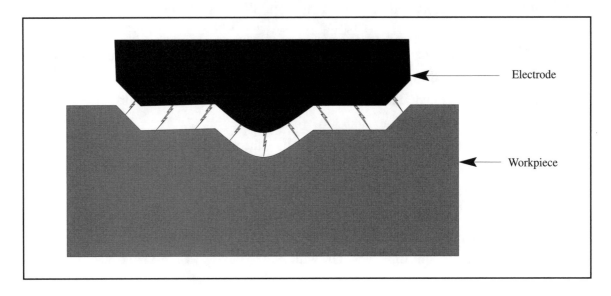

Figure 8:2
Ram EDM process uses a formed electrode to remove material.

Ram EDM Beginnings

Lightning is a form of electrical discharge machining. Its effect can be seen when it strikes the earth. Also, when a screwdriver shorts between a car body and battery, one witnesses how electricity can remove metal.

In 1889, Benjamin Chew Tilghman, of Philadelphia, PA, received a U.S. patent (Patent No. 416,873) entitled, "Cutting Metal By Electricity." This is a portion of the patent:

> My object is to provide a method by which metal objects can not only be severed, but also planed, turned, or shaped in any ordinary way; and I avoid as far as possible heating the metal under treatment except at the point where the cutting action is taking place. This I accomplish by concentrating the electric current upon a path or continuous series of small spots or points adjoining each other, and successively brought under the influence of the current, so that the metal is always heated to the desired degree at the point where it is being operated upon and not elsewhere.

Although Tilghman had developed the concept of electrical discharge machining, spark erosion devices between World War I and World War II were used primarily to remove broken drills and taps. These early machines were very inefficient and difficult to use.

Then two Russian scientists, Boris R. and Natalie I. Lazarenko (husband and wife) made two important improvements. First they developed the R-C relaxation circuit which provided a consistent pulse control. Second they developed a servo control unit which maintained a consistent gap allowing efficient electrical discharges.

These two developments made ram EDM a more dependable means of production. However, the process still had its limitations. For instance, the vacuum tubes used for the direct current circuit could not carry enough current or allow quick switches between "on" and "off" times.

Current and switching problems faded with the introduction of the transistor. Better accuracy and finishes resulted because the solid state device permitted the use of the proper current and switching for "on" and "off."

Today, ram EDM machines have enhanced servo systems, CNC-controls with fuzzy logic, automatic tool changers (Figure 8:3), and capabilities of simultaneous six-axes machining. Ram EDM, along with wire EDM, has revolutionized machining.

Courtesy Agie

Figure 8:3
A CNC Ram EDM With Tool Changer

How Ram EDM Works

Ram EDM uses spark erosion to remove metal. Its DC power supply generates electrical impulses between the workpiece and the electrode. A small gap between the electrode and the workpiece allows a flow of dielectric oil. When sufficient voltage is applied, the dielectric oil ionizes, and controlled sparks melt and vaporize the workpiece.

The pressurized dielectric oil cools the vaporized metal and removes the eroded material from the gap. A filter system cleans the suspended particles from the dielectric oil. The oil goes through a chiller to remove the generated heat from the spark erosion process. This chiller keeps the oil at a constant temperature which aids in machining accuracy. See Figure 8:4

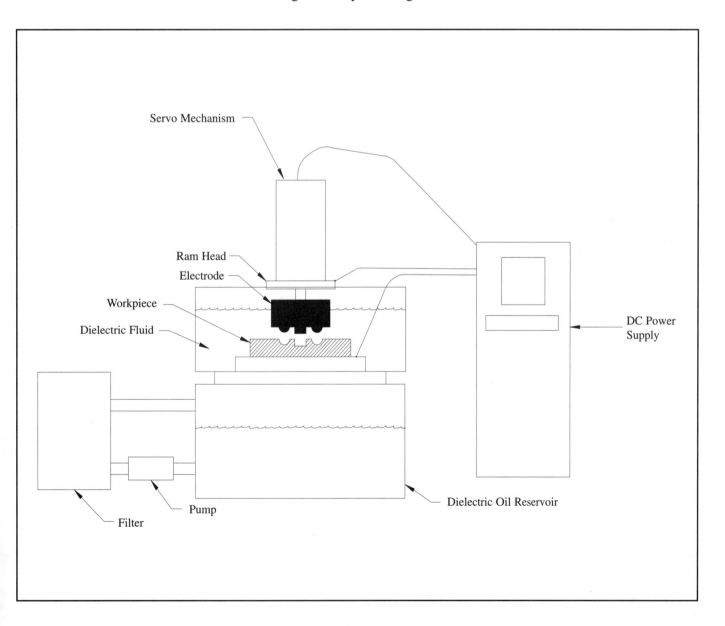

Figure 8:4
The Ram EDM Process

Ram EDM, like wire EDM, is a spark erosion process. However, ram EDM produces the sparks along the surface of a formed electrode, as in Figure 8:5.

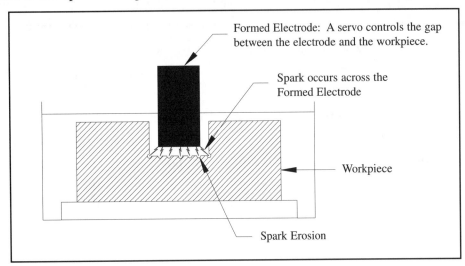

Figure 8:5
Spark Erosion Across the Formed Electrode

A DC servo mechanism maintains the gap between the electrode and the workpiece from .001 to .002″ (.025 to .05 mm). The servo system prevents the electrode from touching the workpiece. If the electrode were to touch the workpiece, it would create a short circuit and no cutting would occur.

The Step-by-Step Ram EDM Process

The DC power supply provides electric current to the electrode and the workpiece. (A positive or negative charge is applied depending upon the desired cutting conditions.) The gap between the electrode and the workpiece is surrounded with dielectric oil. The oil acts as an insulator which allows sufficient current to develop. See Figure 8:6.

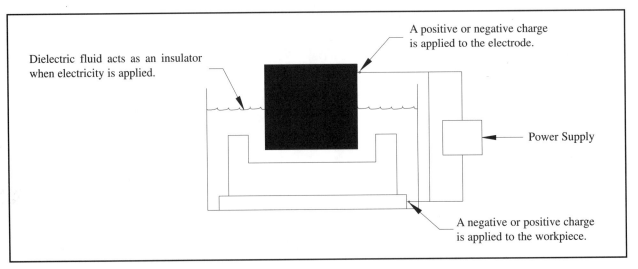

Figure 8:6
Power Supply Provides Volts and Amps

Once sufficient electricity is applied to the electrode and the workpiece, the insulating properties of the dielectric oil break down, as in Figure 8:7. A plasma zone is quickly formed which reaches up to 14,500° to 22,000° F (8,000° to 12,000° C). The heat causes the fluid to ionize and allows sparks of sufficient intensity to melt and vaporize the material. This takes place during the controlled "on time" phase of the power supply.

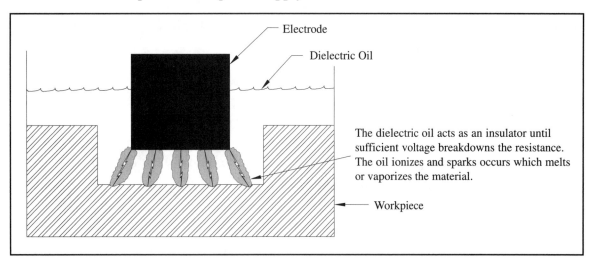

Electrode

Dielectric Oil

The dielectric oil acts as an insulator until sufficient voltage breakdowns the resistance. The oil ionizes and sparks occurs which melts or vaporizes the material.

Workpiece

Figure 8:7
Sparks Causes the Material to Melt and Vaporize

During the "off times," the dielectric oil cools the vaporized material while the pressurized oil removes the EDM chips, as in Figure 8:8. The amount of electricity during the "on time" determines the depth of the workpiece erosion.

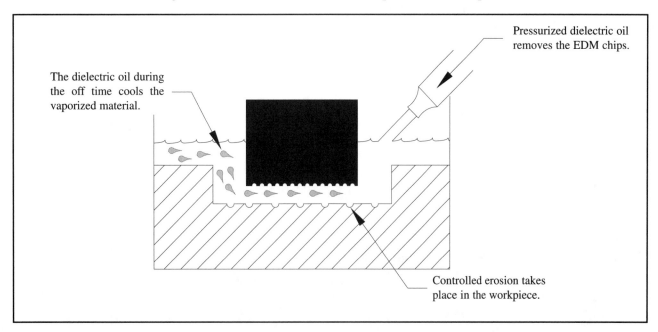

Pressurized dielectric oil removes the EDM chips.

The dielectric oil during the off time cools the vaporized material.

Controlled erosion takes place in the workpiece.

Figure 8:8
Pressurized Dielectric Oil Removes the EDM Chips

Polarity

Polarity refers to the direction of the current flow in relation to the electrode. The polarity can be either positive or negative. (Polarity changes are not used in wire EDM.)

Changing the polarity can have dramatic affects when ram EDMing. Generally, electrodes with positive polarity wear better, while electrodes with negative polarity cut faster. However, some metals do not respond this way. Carbide, titanium, and copper should be cut with negative polarity.

No-Wear

An electrode that wears less than 1% is considered to be in the no-wear cycle. No-wear is achieved when the graphite electrode is in positive polarity and "on times" are long and "off times" are short. However, with the no-wear cycle there is a substantial reduction in metal rate removal.

During the time of no-wear, the electrode will appear silvery showing that the workpiece is actually plating the electrode. During the no-wear cycle there is a danger that nodules will grow on the electrode, thereby changing its shape.

Fuzzy Logic

Some ram machines come equipped with fuzzy logic. Unlike bilevel logic, which recognizes a statement as either true or false, fuzzy logic allows a statement to be partially true or false. Fuzzy logic allows machines to think and react quickly to various machining conditions. These machines can lower or increase power settings to obtain the optimum combination of speed, precision, and finish. Fuzzy logic machines constantly monitor the cut and change power settings to maximize efficiency.

Fumes From the EDM Process

Fumes are emitted during the EDM process; therefore, a proper ventilation system should be installed. Boron carbide, titanium boride, and beryllium are three metals that give off toxic fumes when being EDMed; these metals need to be especially well-vented.

Benefits of Understanding the Process

The better understanding manufacturers have of the EDM process, the better they can use it to reduce costs. The following section discusses how to profit with ram EDM.

9 Profiting With Ram EDM

Users of Ram EDM

There are many operations where ram EDM is the most efficient way to machine parts. In some cases, it is the only way. It is especially useful for blind cavities.

Sometimes numerically controlled mills are used for blind cavities, but when sharp corners, intricate details, or a fine finishes are required, an electrode is produced and the workpiece is EDMed. To form a soft shaped male electrode is often more economical than to mill and polish a female workpiece.

Many industries, particularly the mold industry, use ram EDM extensively. See Figure 9:1. Practically all major mold manufacturers operate their own ram EDM.

Courtesy Agie

Figure 9:1
Single Cavity Mold

Benefits of Ram EDM

A. Different Shapes and Sizes

Ram EDM can machine a wide variety of shapes and sizes as in Figure 9:2. This non-contact method of machining with low-pressure flushing also allows it to produce very thin sections.

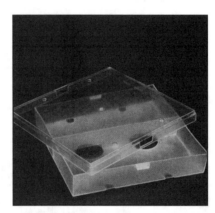

Multi-Cavity Mold for Plastic Containers

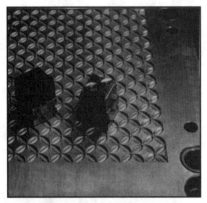

Mold for Rubber Mat   **Mold for Motor Rotor Cooling Blades**

Mold for Glass Stems   Courtesy Charmille

Figure 9:2
Examples of Molded Shapes Produced With Ram EDM

B. Accuracy and Finishes

Depending on the accuracy of the electrode, tolerances of up to +/- .0001″ (.0025 mm) can be held. Furthermore, if the correct amount of current is used, very fine finishes can be obtained.

Certain machines can produce a mirror-type finish. Machines capable of producing mirror finishes eliminate the laborious method of polishing the cavities.

C. Workpiece Hardness Not a Factor

Workpiece hardness has no effect on cutting. Therefore hardened parts can be easily machined.

D. EDMing Threads Into Hardened Parts

Ram EDM is capable of machining threads into hardened parts, difficult-to-machine alloys, and even carbide. CNC machines are capable of doing this by orbiting a thread electrode. But units are available that can be mounted on any ram EDM machine to cut threads from #5-40 to 3/4″-10. There are also special units that will cut threads as small as #0-80, threads as large as several inches, and pipe threads. See Figure 9:3.

Courtesy Custom EDM

Figure 9:3
Ram EDM Tapping Head for Conventional EDMs

Parts for Ram EDM

A. Molds

Ram EDM produces molds, from miniature toys to large injected plastic molded parts for automobiles. Plastic is injected into these preformed molds and then cooled.

Courtesy Charmille

Figure 9:4
A Graphite Electrode and the Molded Part

B. Blind Keyways

Ram EDM can easily cut blind keyways as in Figure 9:5. Wire EDM is usually more cost effective when keyways pass through the part.

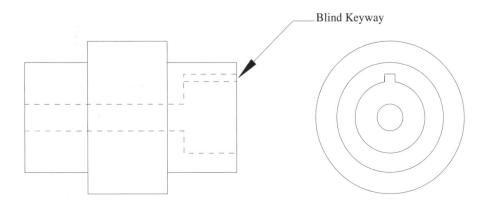

Blind Keyway

Figure 9:5
Blind Keyway

C. Internal Splines

When internal splines do not go through the part as in Figure 9:6, then ram EDM is used to machine the splines.

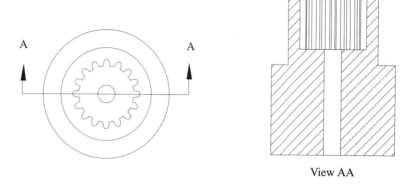

View AA

Figure 9:6
Internal Splines

D. Hexes for Special Bolts and Parts

Ram EDM is ideal to machine special bolts and parts with blind cavities, such as hexes as shown in Figure 9:7.

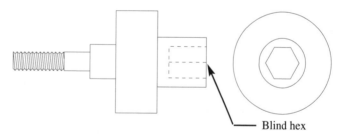

Blind hex

Figure 9:7
Hexes for Special Bolts and Parts

E. Helical Gear Machining

Orbiting machines can machine helical gears, as seen in Figure 9:8.

Courtesy Mitsubishi

Figure 9:8
Helical Gear Machining

Materials For Ram EDM

Any electrically-conductive material can be cut with ram EDM, such as: tool steels, cold and hot rolled steel, stainless steels, inconel, hastalloy, stellite, aluminum, copper, brass, titanium, and carbide.

Speeding the Mold Processing

Mold makers often seek ways to speed up removing molded material. When certain mold areas take longer to cool than other areas, cycle times must be lengthened. Adding more water lines is not always feasible due to the configuration of the mold.

Processing speeds may be increased by placing a high thermal conductivity copper alloy, like Ampco alloy 940, into areas requiring faster cooling. Using such copper alloys can reduce the cycle time from 20% to 30%, since this material disperses heat six times faster than steel.

To EDM these copper alloys, a high grade graphite with negative polarity should be used. When machining this high thermal material, it is recommended that 50 to 60 amps be used.

Another method to cool the mold quickly without substantially changing it is to replace steel core pins with copper alloy pins.

EDMing Carbide

Carbide ranges from high cobalt (16%), which is a low wear, high shock grade, to low cobalt (6%), which is a high wear, low shock grade. Since only the cobalt in carbide conducts electricity, the carbide does not EDM as rapidly as steel . Therefore, the higher the percentage of cobalt, the faster the carbide can be EDMed.

Proper Procedures for Ram EDM

Many parts would be impossible to be machined without ram EDM. It is important to learn the proper procedures to maximize the benefits of this process, for by learning the proper use of Ram EDM, it can dramatically reduce operating costs. The next few chapters will discuss the proper procedures for ram EDM.

10 Ram EDM Electrodes and Finishing

Electrodes

Electrode selection and machining are important factors in operating ram EDM.

A. Function of the Electrode

The purpose of an electrode is to transmit the electrical charges and to erode the workpiece to a desired shape. Different electrode materials greatly affect machining. Some will remove metal efficiently but have great wear; other electrode materials will have slight wear but remove metal slowly.

B. Electrode Selection

When selecting an electrode and its fabrication, these factors need to be evaluated:
1. Cost of electrode material.
2. Ease or difficulty of making an electrode.
3. Type of finish desired.
4. Amount of electrode wear.
5. Number of electrodes required to finish the job.
6. Type of electrode best suited for the work.
7. Number of flushing holes, if required for the electrode.

C. Type of Electrode Materials

Electrodes fall into two main groups: metallic and graphite. There are five commonly used electrodes: brass, copper, tungsten, zinc, and graphite. In addition, some electrode materials are combined with other metals in order to cut more efficiently: brass and zinc, copper and tellurium, copper and tungsten, tungsten and silver, and graphite and copper.

Initially, brass was used as an electrode in spite of its high wear. Later, operators increased wear ratio by using copper and its alloys. The problem with copper is that its melting point is approximately 1085° C.; whereas the spark temperature in the gap exceeds 3800° C. Copper's low melting point often causes too high a wear rate in relation to its metal removal.

Studies show that graphite electrodes have greater rate of metal removal in relation to its wear. Graphite does not melt in the spark gap; rather, at approximately 6062° F (3350° C), it changes from a solid to a gas. Because of graphite's relatively high resistance to heat in the spark gap (as compared to copper), for most jobs it is a more efficient electrode material. See Figure 10:1. Tungsten has a melting point similar to graphite, but tungsten is extremely difficult to machine. Tungsten is used as "preforms," usually as tubing and rods for holes and small hole drilling.

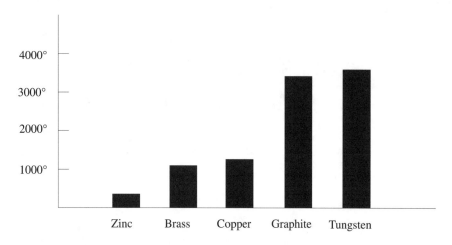

Figure 10:1
Electrode Melting Points

Metallic electrodes usually work best for EDMing materials which have low melting points as aluminum, copper, and brass. As for steel and its alloys, graphite is preferred. The general rule is:

> Metallic electrodes for low temperature alloys.
> Graphite electrodes for high temperature alloys.

However, exceptions exist. For instance, despite higher melting points for tungsten, cobalt, and molybdenum, metallic electrodes like copper are recommended due to the higher frequencies needed to EDM these materials.

Copper has a distinct advantage over graphite because it performs better in "discharge-dressing." During unsupervised CNC cutting, the copper electrode can be sized automatically by using a sizing plate. The copper electrode can then be reused for a finishing cut or used to produce another part.

D. Galvano Process for Metallic Electrodes

Sometimes large solid electrodes are too heavy for the servo and too costly to fabricate. In such cases the Galvano process can be used to fabricate the mold. A mold is electrolytically deposited with copper up to .200″ (5 mm) thick. The inside of the copper shell is partially filled with an epoxy, and wires are attached to the copper electrode. The formed electrode is then mounted on the EDM machine.

E. Custom Molded Metallic Electrodes

Where multiple electrodes are constantly required, a 70/30 mixture of tungsten and copper powder is pressure molded and sintered in a furnace. This process can produce close tolerance electrodes.

F. Graphite Electrodes

In America, approximately 85 percent of the electrodes used are graphite. Graphite machines and grinds easily compared to metal electrodes. Burrs usually occur when machining metal electrodes; however, burrs are absent when machining graphite. Copper tends to clog grinding wheels. To avoid wheel clogging, some use an open grain wheel and beeswax, or a similar product.

However, graphite has a major problem: it is "dirty." Unlike metal when it is machined, graphite does not create chips—it creates black dust. If graphite dust is not removed while being machined, it will blanket the shop. Although certain graphites are used for lubricants, the graphite in electrodes is synthetic and very abrasive. Getting graphite into the machine ways can cause premature wear. Because of the abrasive characteristics of graphite, machinists are advised to use carbide cutting tools. When grinding graphite electrodes, they should use a vacuum system. See Figure 10:2.

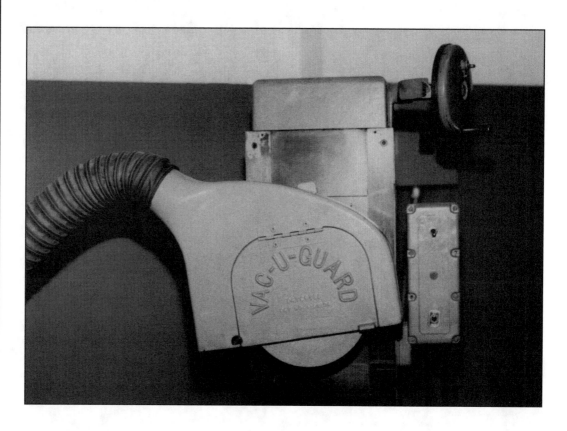

Figure 10:2
A Surface Grinder Equipped With a Vacuum System for Grinding Graphite

A vacuum system also can be installed when milling graphite. Some mills use a liquid shield around the cutter to remove graphite dust. There are also special designed, totally enclosed milling machines that are used to machine graphite.

Graphite is porous, and liquids can penetrate and introduce problem-causing impurities. The larger the graphite grain structure, the greater the danger for impurities. However, dense graphite, even after being soaked in fluid for several hours, shows little fluid penetration.

One way to remove impurities is to put the electrode in an oven for one hour at 250° F (121° C). Electrodes can also be air dried. It is recommended that graphite electrodes should never be placed in a microwave oven.

If porous electrodes are used, they should contain no moisture. Trapped moisture can create steam when cutting and damage the electrode.

When machining graphite, it tends to chip when exiting a cut. To prevent chipping, machinists should use sharp tools, low feed rates, and a positive rake. A method to prevent chipping is to make a precut into the graphite where the cutting tool will exit.

Different grades and porosities of graphite are shown in Figure 10:3

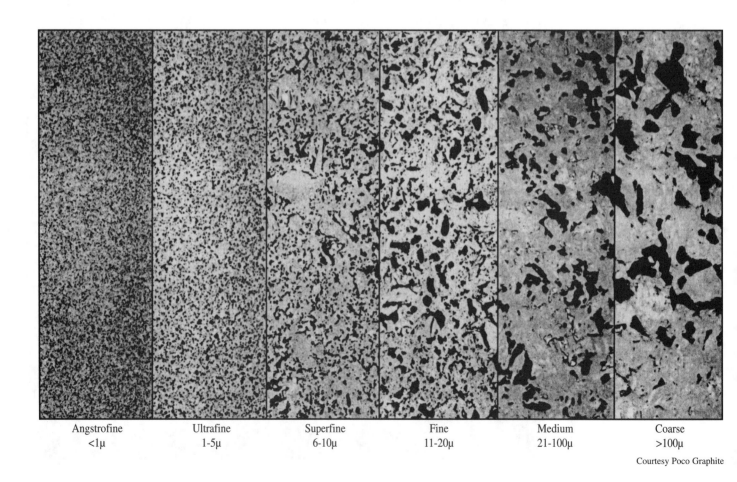

| Angstrofine | Ultrafine | Superfine | Fine | Medium | Coarse |
| <1μ | 1-5μ | 6-10μ | 11-20μ | 21-100μ | >100μ |

Courtesy Poco Graphite

Figure 10:3
Graphite Grain Size Magnified 100 X

G. Determining Factors for Choosing the Proper Graphite

Grain size and density of graphite determine its cost and cutting efficiency. See Figure 10:4.

Angstrofine—Used to where extremely fine detail and very smooth finishes are required.

Ultrafine—Used to attain strength, electrode detail, good wear and fine surface finish are necessary.

Superfine—Used in large molds where detail is maintained and speed is important.

Fine—Used in very large cavities where detail and finish are not critical.

Courtesy Poco Graphite

Figure 10:4
Typical Electrode Shapes for Various Classifications of Graphite

The general rule for determining graphite is:

Choose a finer grain size graphite for fine detail, good finish, and high wear resistance.

Choose a less costly, coarser electrode when there is no concern for small detail or fine finish.

H. Electrode Wear

Except in the no wear cycle, electrodes have considerable wear. If the portion of the electrode that did not wear retains it shape, the electrode can be redressed and reused. For example: A long hex graphite is machined for blind hex cavities. When the lower portion of the hex electrode wears, its worn portion is removed and the electrode is reused.

On some formed electrodes, an electrode cannot be remachined. In such cases, sufficient electrodes need to be fabricated.

Heaviest electrode wear appears in the corners. This wear occurs because the electrode corner must EDM a larger area than other surfaces. See Figure 10:5.

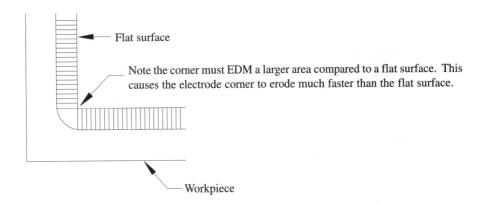

Figure 10:5
Corner Electrode Wear

I. Abrading Graphite Electrodes

The abrading process is an efficient method of producing complex and large electrodes for production and redressing purposes. A pattern is first made for the desired shape. Then an epoxy inverted form is made from the pattern and charged with a carbide grit coating. This carbide-grit form becomes the abrading tool. See Figure 10:6.

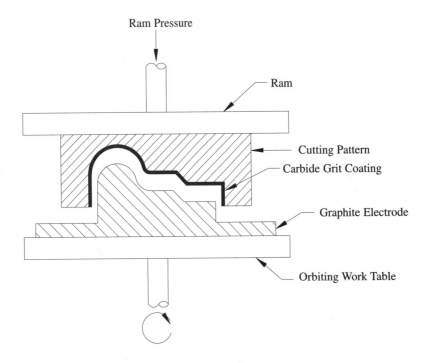

Figure 10:6
Abrading Graphite Electrodes

The machine orbits from .020″ to .200″ (.51 mm to 5 mm). As the machine vibrates in a circular motion within a bath of oil, the impregnated pattern forms the graphite electrode. See Figure 10:7.

Courtesy Hausermann

Figure 10:7
Abrading Machine

The abrading tool produces a very fine finish on the electrode. Multiple electrodes can be produced from the same pattern without any secondary benchwork. See Figure 10:8. This process is used for large electrodes with many details, such as crankshaft forging dies and transmission housing molds. See Figure 10:9. Abrading is further discussed in Chapter 14, "Abrasive Flow, Thermal Energy Deburring, and Ultrasonic Machining."

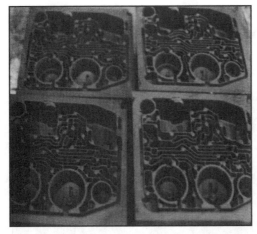

Courtesy Hausermann   Courtesy Hausermann

Figure 10:8   **Figure 10:9**
Abraded Valve Body Electrodes for   **Large Abraded Electrode for Plastic Mold**
Automatic Transmission   **for Bumper Fascia**

I. Ultrasonic Machining for Graphite Electrodes

As in abrading, ultrasonic machining also cuts by vibration. It uses a metal form tool and an abrasive slurry flow between the form tool and the electrode. The electrode is formed as the workpiece vibrates. This process is predominantly used for shallow cavities, such as coining and embossing dies. Ultrasonic machining is more fully described in Chapter 14, "Abrasive Flow, Thermal Energy Deburring, and Ultrasonic Machining."

J. Wire EDMing Metallic and Graphite Electrodes

Some believe wire EDMing metallic electrodes is efficient, whereas wire EDMing graphite electrodes is inefficient. However, in recent years the cutting speeds of wire EDM have increased, making it economical for cutting graphite electrodes.

In addition, when electrodes containing fine details are wire EDMed, the fine details add no significant costs to electrode fabrication. Also, the dust problem associated with machining graphite electrodes is eliminated because deionized water in wire EDM washes the eroded particles away.

The densely-structured Angstrofine graphite cuts nearly twice as fast as all other graphites. Zinc coated wires have also increased the speed of wire EDMing graphite electrodes. Some studies show that using zinc coated wires have increased cutting speeds up to 50 percent.

K. Electrode Overcut

The EDMed cavity will always be larger than the electrode. The difference between the electrode and the workpiece gap is called the "overcut," or "overburn," as shown in Figure 10:10. The amount of overcut will vary according to the amount of current, "on times," type of electrode, and workpiece material.

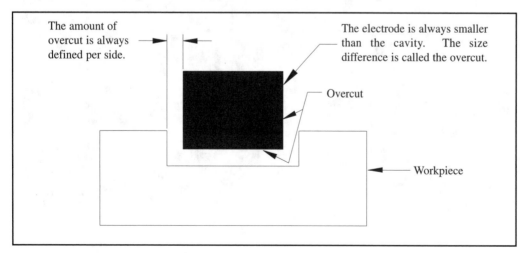

Figure 10:10
The Overcut

The primary factor affecting the overcut is the amount of electrical current in the gap. The overcut is always measured per side. Overcuts can range from a low of .0008″ (.020 mm) to a high of .025″ (.63 mm). The high overcuts are the results of cutting with high amperages. Most manufacturers have charts showing the amount of overcut operators can expect with certain power settings.

During a roughing cut, greater current is applied to the electrode, causing a greater overcut. A finishing cut, however, uses less current and produces a much smaller overcut.

Given the same power settings and material, the overcut remains constant. For this reason, tolerances to +/- .0001 (.0025 mm) can be achieved with ram EDM. However, when such tolerances are called for, the cost increases because machining time increases.

Recast and Heat-Affected Zone

The EDM process creates three types of surfaces. The top surface contains a thin layer of spattered material that has been formed from the molten metal and the small amounts of electrode material. This surface layer of spattered EDM residue is easily removed.

Underneath the spattered material is the recast (white) layer. When the current from the EDM process melts the material, it heats up the underlying surface and alters the metallurgical structure.

This recast layer is formed because some of the molten metal has not been expelled and has instead been rapidly quenched by the dielectric oil. Depending on the material, the recast layer surface can be altered to such an extent that it becomes a hardened brittle surface where microcracks can appear. This layer can be reduced substantially by finishing operations.

The third layer is the heat-affected zone. This area is affected by the amount of current applied in the roughing and finishing operations. The material has been heated but not melted as in the recast layer. The heat-affected zone may alter the performance of the material.

There can be significant differences between wire and ram EDM heat-affected zones. When roughing with ram EDM, much more energy can be supplied then with wire EDM. This greatly increases the heat-affected zone with Ram EDM. On thin webs it can create serious problems because the material will be heat treated and quenched in the dielectric oil. This can cause thin webs to become brittle.

When dielectric oil is heated, the hydrocarbon in the oil breaks down and creates an enriched carbon area in the cutting zone. This carbon becomes impregnated into the surface and alters the parent material. Often this surface becomes hard and makes polishing more difficult. To avoid heat problems when EDMing thin webs, parts should be premachined and EDMed with lower power settings.

Today's newer power supplies create about half the depth of heat-affected zones as older machines. This shallower depth reduces the need for removing more material to reach base metal.

The depth of the altered metal zone changes according to the amount of current applied, as shown in Figure 10:11. A careful finishing operation can greatly reduce these three layers of the heat-affected zones.

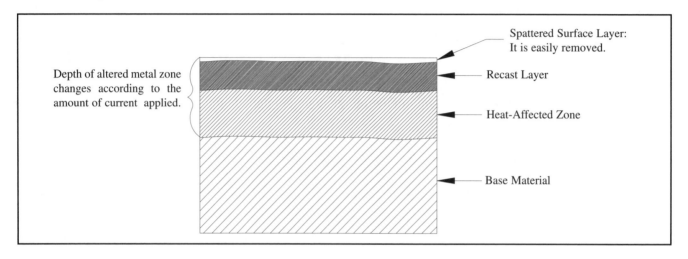

Figure 10:11
Metal Zones Altered by EDM

Finishing

Knowing the principle of the overcut is important to understand the resulting surface finish. When high current is applied to the workpiece, it produces large sparks and large workpiece craters. This results in a rough finish, as illustrated in Figure 10:12

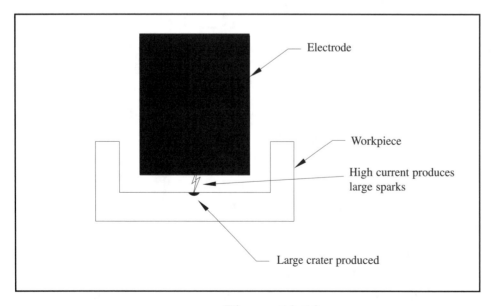

Figure 10:12
Roughing Cut Produces a Coarse Finish

When a slight amount of current is applied to the workpiece, small sparks are produced which create small craters. Applying low current slows the machining process, but it produces a fine finish, as shown in Figure 10:13.

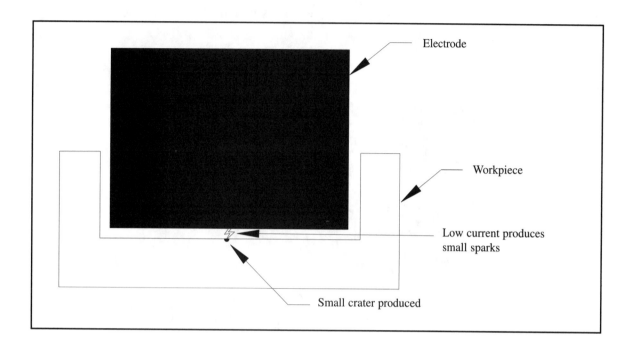

Figure 10:13
Finishing Cut With Low Current Produces a Fine Finish

When a very small amount of current is applied (short on times and low peak current) to the surface of the workpiece, machines are capable of producing mirror-like finishes. Machines equipped with orbiting abilities can also help to produce a fine finish by orbiting the electrode. Certain orbiting machines can be programmed so that the current is gradually reduced until a mirror-like finish occurs.

The workpiece finish will be a mirror image of the electrode. If the electrode is imperfect or pitted, the finish will be imperfect or pitted. A coarse electrode produces a coarse finish. The finer the electrode grain structure, the finer the finish.

Mirror Finishing

Advances in the controls and the dielectric fluid have dramatical improved surface finish. Some machines use a specially formulated dielectric fluid for finishing operations that produces mirror finishes of less than 1.5 Rmax p17μm. Some machines contain two dielectric fluid tanks, one for conventional roughing and semi-finishing and the other for producing mirror finishes.

Manufacturers have discovered that after adding silicon, graphite, or aluminum powder to the dielectric fluid, excellent surface finishes are produced. What transpires is the electrical discharges from the electrode to not first strike the workpiece, but strike the silicon or other particles and generate micro discharges. These micro electrical discharges result in craters so small that they produce a mirror finish. See Figure 10:14.

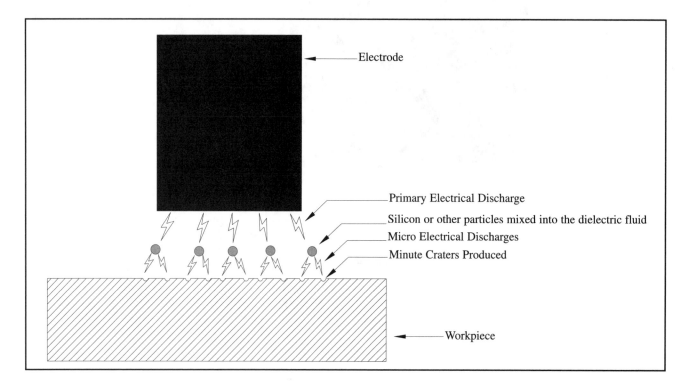

Figure 10:14
Mirror Finishes

The specially formulated dielectric fluid allows the gap distance between the electrode and the workpiece to increase from .0008″ to .004″ (.020 mm to .1 mm) and more. This larger gap greatly improves the flushing and results in a much more stable cut. Also, the current is distributed more evenly, greater surface areas can be machined, and higher spark energy can be used. Machining speeds with this process for finishing have increased over 600%.

Micro Machining

Micro machining with EDM is being done with electrodes as small as .0004″ (.01 mm). Micro-machining uses specialized machines using low power and equipped with microscopes for viewing and inspection.

Micro stamping is being explored at the University of Tokyo with punches as small as .0012″ (.03 mm). They use a wire electrode to EDM the micro punch. The front end of the punch is used to EDM the die section. After the die section is EDMed, the front end of the punch is removed by EDMing the thin section off. They use the micro punch to stamp .002″ (.05 mm) phosphor bronze material in the EDM machine. Obviously, this procedure is not for volume production. See Figures 10:15 - 10:18.

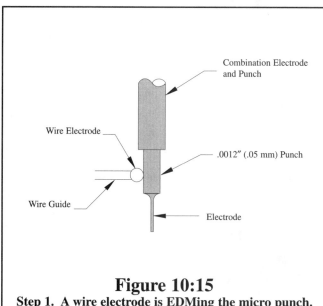

Figure 10:15
Step 1. A wire electrode is EDMing the micro punch.

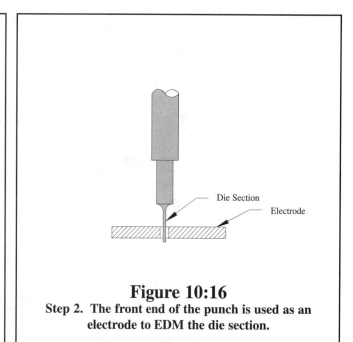

Figure 10:16
Step 2. The front end of the punch is used as an electrode to EDM the die section.

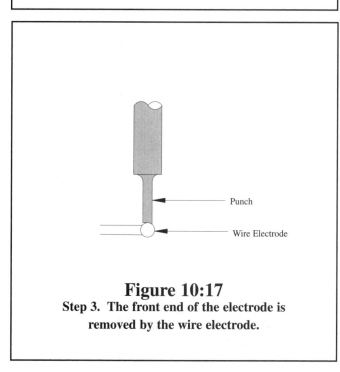

Figure 10:17
Step 3. The front end of the electrode is removed by the wire electrode.

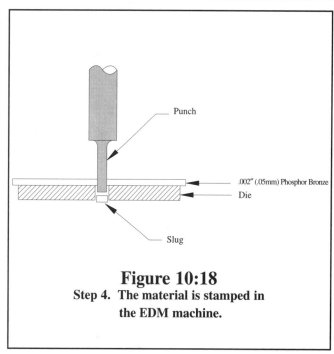

Figure 10:18
Step 4. The material is stamped in the EDM machine.

Figure 10:15-18
Micro Machining

For small holes and slots, lasers have been the instrument of choice. However, sometimes the edges of the laser holes or slots have poor edge definition. With micro EDM the edges of the holes and slots are square. This capability is particularly useful for items as optical apertures and guides, ink-jet printer nozzles, audio-visual components, and computer peripherals. See Figure 10:19.

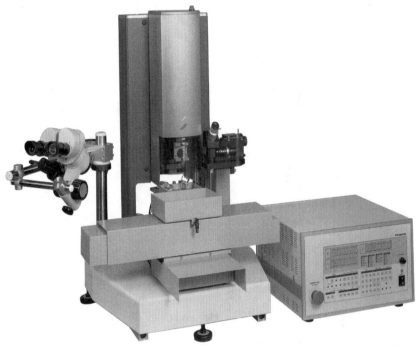

Courtesy Panasonic

Figure 10:19
Micro EDM Machine

Ram EDM has many exciting possibilities. The next section covers the function of the dielectric oil and the various ways of flushing.

11 Dielectric Oil and Flushing for Ram EDM

Dielectric Oil

Ram EDM uses oil for its dielectric fluid. Dielectric oil performs three important functions for ram EDM, see Figure 11:1.

1. The oil forms a dielectric barrier for the spark between the workpiece and the electrode.

2. The fluid cools the eroded particles between the workpiece and the electrode.

3. The pressurized oil flushes out the eroded gap particles and removes the particles from the oil by causing the oil to pass through a filter system.

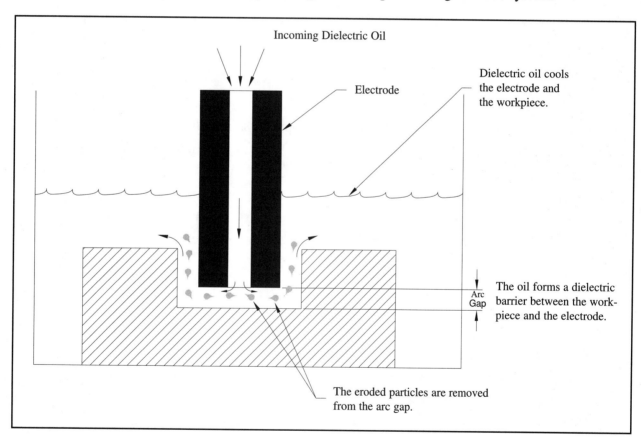

Figure 11:1
Functions of the Dielectric Oil

Many manufacturers produce many types of dielectric oil; the best way to determine the type of oil needed for a particular machine is to ask the machine manufacturer for its recommendations. It is important to get oil which is specifically produced for ram EDM.

Coolant System

EDM creates sparks in the gap with sufficient energy to melt the material. The resulting heat is transferred into the oil. Oil loses its efficiency when it reaches 100° F (38° C). Controlling this heat is essential to ensure accuracy and efficient cutting. Therefore, it is best to have a coolant system to maintain a proper temperature.

Flash Point

Oil will ignite at certain temperatures. The ignition temperature is called "flash point." This is especially important when doing heavy cutting, because the oil may get so hot that it reaches its flash point. Even though some oils have a flash point of 200° F (93° C) and higher, it is unsafe to use oil over 165° F (74° C). Precautions need to be taken to prevent the oil from reaching its flash point. Some machines are equipped with a fire suppression system that is controlled by an infrared scanner.

Flushing

A. Proper Flushing

The most important factor in EDM is to have proper flushing. There is an old saying among EDMers: "There are three rules for successful EDMing: flushing, flushing, and flushing."

Flushing is important because eroded particles must be removed from the gap for efficient cutting. Flushing also brings fresh dielectric oil into the gap and cools the electrode and the workpiece. The deeper the cavity, the greater the difficulty for proper flushing.

Improper flushing causes erratic cutting. This in turn increases machining time. Under certain machining conditions, the eroded particles attach themselves to the workpiece. This prevents the electrode from cutting efficiently. It is then necessary to remove the attached particles by cleaning the workpiece.

The danger of arcing in the gap also exists when the eroded particles have not been sufficiently removed. Arcing occurs when a portion of the cavity contains too many eroded particles and the electric current passes through the accumulated particles. This arcing causes an unwanted cavity or cavities which can destroy the workpiece. Arcing is most likely to occur during the finishing operation because of the small gap that is required for finishing. New power supplies have been developed to reduce this danger.

B. Volume, Not Pressure

Proper flushing depends on the volume of oil being flushed into the gap, rather than the flushing pressure. High flushing pressure can also cause excessive electrode wear by making the eroded particles bounce around in the cavity. Generally, the ideal flushing pressure is between 3 to 5 psi. (.2 to .33 bars).

Efficient flushing requires a balance between volume and pressure. Roughing operations, where there is a much larger arc gap, requires high volume and low pressure for the proper oil flow. Finishing operations, where there is a small arc gap, requires higher pressure to ensure proper oil flow.

Often flushing is not a problem in a roughing cut because there is a sufficient gap for the coolant to flow. Flushing problems usually occur during finishing operations. The smaller gap makes it more difficult to achieve the proper oil flow to remove the eroded particles.

C. Types of Flushing

There are four types of flushing: pressure, suction, external, and pulse flushing. Each job needs to be evaluated to choose the best flushing method.

1. Pressure Flushing

Pressure flushing, also called injection flushing, is the most common and preferred method for flushing. One great advantage of pressure flushing is that the operator can visually see the amount of oil that is being used for flushing. With pressure gauges, this method of flushing is simple to learn and use. Pressure flushing may be performed in two ways: through the electrode (Figure 11:2) or through the workpiece.

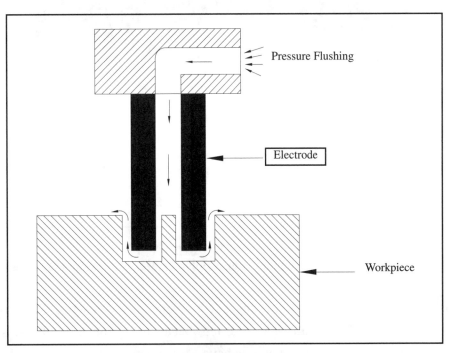

Figure 11:2
Pressure Flushing Through the Electrode

a. Pressure Flushing Through the Electrode

A problem with flushing through the electrode is when a stud or spike remains from the electrode flushing hole. If the stud gets too long, it can hinder proper flushing.

Occasionally, thin studs from soft metal weaken and touch the electrode and cause a short. These studs need to be removed either by hand, by a portable hand grinder, or with another electrode. Studs in hardened metal can be snapped off easily with needle nose pliers. If the machine has orbiting capabilities, the stud can be removed by orbiting.

Small holes are often drilled into large electrodes to aid in flushing, as in Figure 11:3. The placement of these holes is critical to ensure that flushing occurs over the entire cutting area. However, for certain applications, today's newer adaptive controls with fuzzy logic have reduced the need for flushing holes.

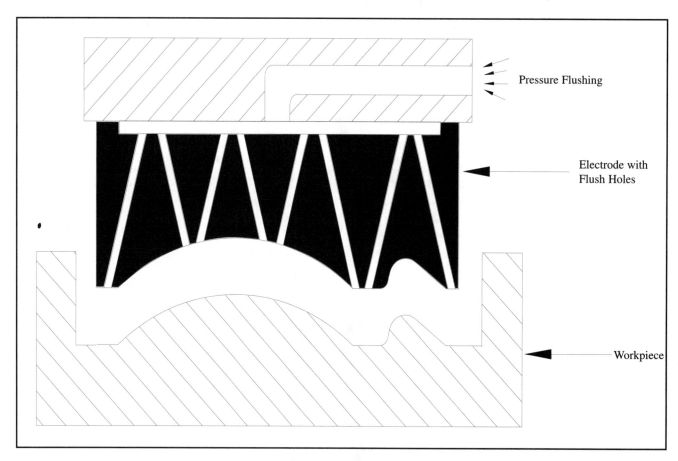

Figure 11:3
Multiple Flush Holes Drilled through the Electrode

Flushing holes are drilled on an angle to prevent long studs from developing. See Figure 11:4. The studs get EDMed because of the angled studs. A disadvantage of angled holes is that they have a tendency to prevent proper flushing to occur by directing oil away from needed areas.

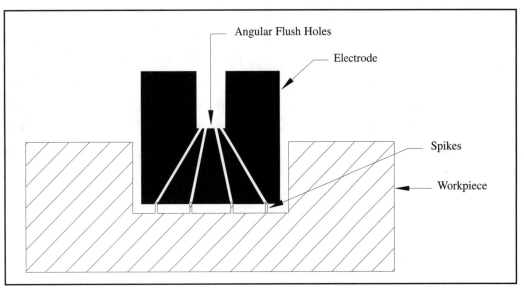

Figure 11:4
Tapering the Flush Holes Prevents Long Studs

When drilling stud holes, machinists often cover the roughing electrode with tape before drilling. They cover the holes with tape to prevent from drilling flushing holes in the same location for the finishing electrode. Holes are drilled in different locations so that previous studs can be removed with the finishing electrode.

When electrodes are the same on both ends, the drilled flushing holes should be offset. By rotating the electrode 180 degrees, the previous studs are EDMed. See Figure 11:5.

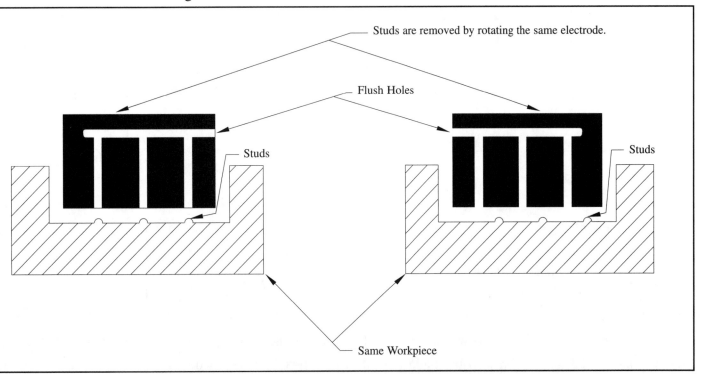

Figure 11:5
Rotating an Electrode 180 Degrees to Remove Studs

Stepped electrodes are sometimes used. If the flushing holes are uncovered in these stepped electrodes, the dielectric oil will escape and prevent flushing in the other holes. To insure the flushing process, a .120″ (3.05 mm) hole is drilled, and a 1/8″ (3.18 mm) diameter silicon tubing (commonly found in aquarium supply stores) is pressed into the electrode to seal the exposed hole. When that portion of the electrode is sealed, the tubing is pulled back so flushing can escape from the hole, as depicted in Figure 11:6.

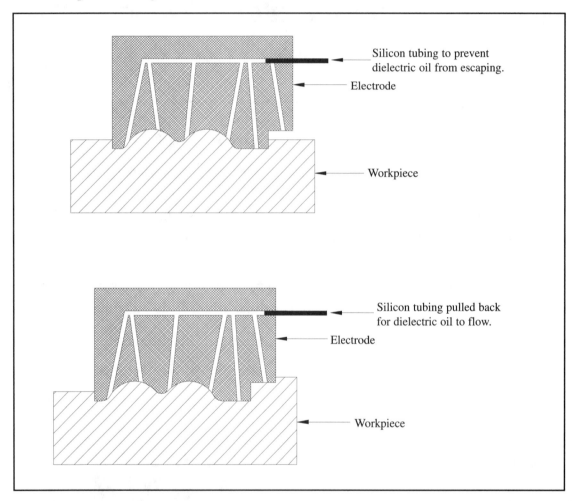

Figure 11:6
Flushing Holes for Stepped Cavities

Sometimes, blind, deep, narrow slots need to be EDMed. If at all possible, a plate should be placed underneath and the more efficient wire EDM process should be used. However, if wire EDM is not possible, one can glue copper tubing onto the sides of a thin electrode, as illustrated in Figure 11:7. The copper tubing will help flush the eroded particles out of the narrow cavity. The glue must be electrically conductive. There are a number of such glues; one recommended glue is Eastman's 910. Glue can also be made from file shavings of fine powdered graphite mixed with super glue. After the copper tubing is glued, the entire electrode should be checked to make certain it is conductive.

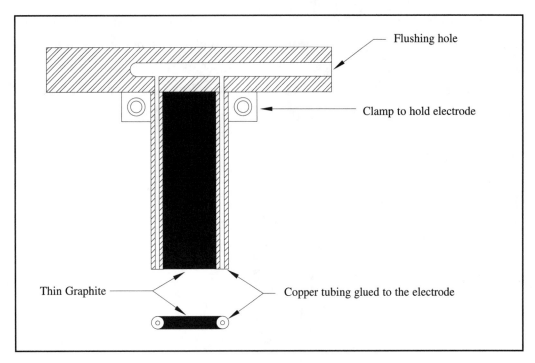

Figure 11:7
EDMing Deep Narrow Slots

b. Pressure Flushing Through the Workpiece

Pressure flushing can also be done by forcing the dielectric fluid through a workpiece mounted over a flushing pot. See Figure 11:8. This method eliminates the need for holes in the electrode.

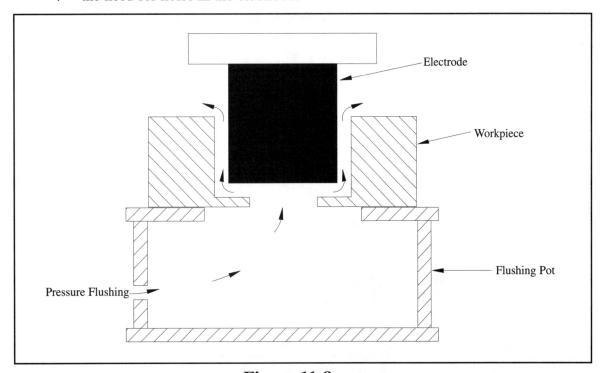

Figure 11:8
Pressure Flushing Through the Workpiece

With pressure flushing, there is the danger of a secondary discharge. Since electricity takes the path of least resistance, secondary discharge machining can occur as the eroded particles pass between the walls of the electrode and the workpiece, as presented in Figure 11:9. This secondary discharge can cause side wall tapering. Suction flushing can prevent side wall tapering.

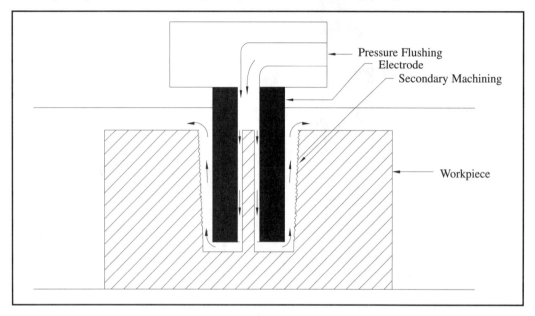

Figure 11:9
Pressure Flushing May Cause Secondary Machining

2. Suction Flushing

Suction or vacuum flushing can be used to remove eroded gap particles. Suction flushing can be done through the electrode as in Figure 11:10, or through the workpiece, as in Figure 11:11.

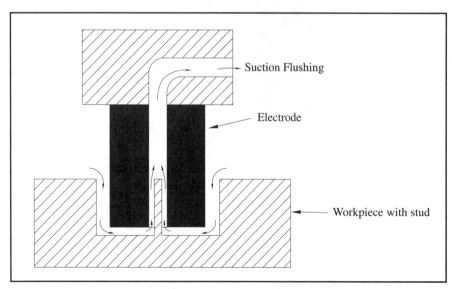

Figure 11:10
Suction Flushing Through the Electrode

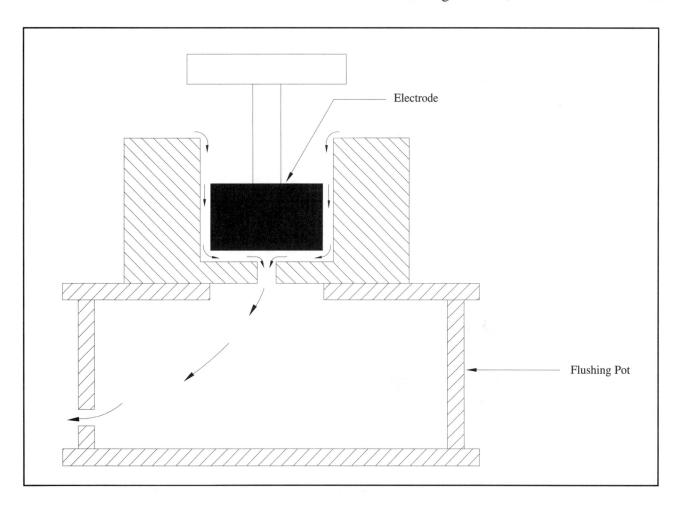

Figure 11:11
Suction Flushing Through the Workpiece

Suction flushing minimizes secondary discharge and wall tapering. Suction flushing sucks oil from the worktank, not from the clean filtered oil as in pressure flushing. For suction cutting, efficient cutting is best accomplished when the work tank oil is clean.

A disadvantage of suction flushing is that there is no visible oil stream as with pressure flushing. Also, gauge readings are not always reliable regarding the actual flushing pressure in the gap.

A danger of suction flushing is that gases may not be sufficiently removed, this can cause the electrode to explode. In addition, the created vacuum can be so great that the electrode can be pulled from its mount, or the workpiece pulled from the magnetic chuck.

3. Combined Pressure and Suction Flushing

Pressure and suction flushing can be combined. They are often used for molds with complex shapes. This combination method allows gases and eroded particles in convex shapes to leave the area and permit circulation for proper machining.

4. Jet Flushing

Jet or side flushing is done by tubes or flushing nozzles which direct the dielectric fluid into the gap, as shown in Figure 11:12.

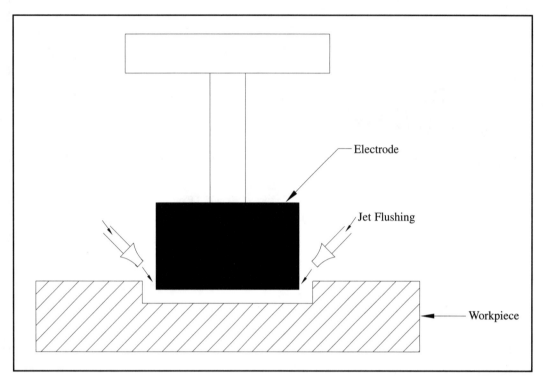

Figure 11:12
Jet Flushing Using Multiple Flushing Nozzles

Although jet flushing is a convenient method of flushing, and sometimes the only choice, it is also the most ineffective way to remove eroded gap particles. The danger of not removing the gap particles is that DC arcing can occur. When placing the nozzles for jet flushing, the fluid must be directed so it can remove the eroded particles from the gap.

One advantage of jet flushing, however, is that it leaves no stud. For shallow cuts, this is an effective method. But as depth increases, external flushing decreases in its effectiveness. Pulse flushing is usually used along with jet flushing.

5. Pulse Flushing

Three types of pulse flushing are:

a. Vertical flushing: the electrode moves up and down.

b. Rotary flushing: the electrode rotates.

c. Orbiting flushing: the electrode orbits.

a. Vertical Flushing

In vertical flushing, the electrode moves up and down periodically in the cavity. This up and down motion causes a pumping action which draws in fresh dielectric oil. See Figure 11:13. Along with vertical flushing, additional external flushing nozzles can be used to help to remove the eroded particles.

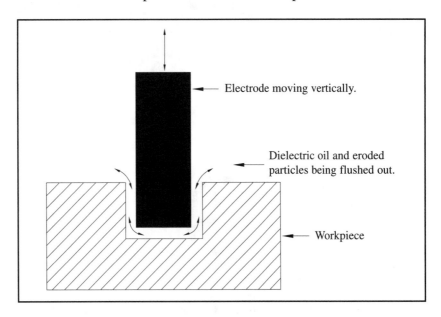

Figure 11:13
Vertical Flushing: The Electrode Moves Up and Down

b. Rotary Flushing

In rotary flushing, the electrode rotates in the cavity as in Figure 11:14.

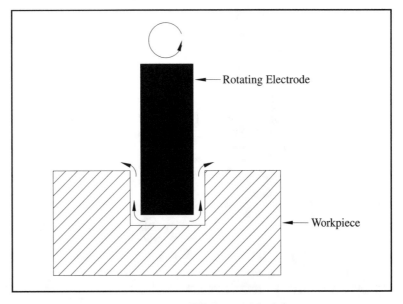

Figure 11:14
Rotary Flushing: The Electrode Rotates

For small round electrodes, manufacturers make multiple cavities in these electrodes to aid in flushing. This is a very efficient method of producing holes without a stud. See Figure 11:15 .

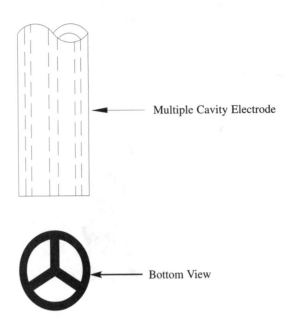

Multiple Cavity Electrode

Bottom View

Figure 11:15
Electrode With Multiple Cavities for Rotary EDMing

c. Orbiting Flushing

Orbiting an electrode in a cavity allows the electrode to mechanically force the eroded particle from the cavity, as pictured in Figure 11:16.

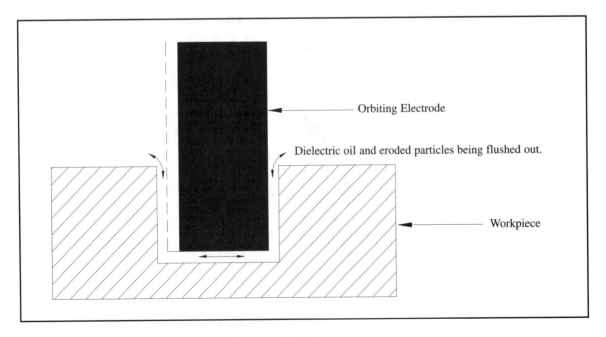

Orbiting Electrode

Dielectric oil and eroded particles being flushed out.

Workpiece

Figure 11:16
Orbiting Flushing: The Electrode Orbits in the Workpiece

Orbiting flushing is the most efficient method for cutting. Furthermore, if the orbiting is larger than the radius of the flushing holes in the electrode, it will produce no studs.

Filtration System

In order to insure proper cutting, a filtration system needs to be maintained that adequately removes the eroded particles from the dielectric oil. Improperly filtered oil will send oil with eroded particles into the gap which will hinder effective cutting.

The Challenge of New Procedures

Reducing costs should always be on the minds of manufacturers. One of the best ways to reduce costs is to understand the process and search for new procedures. The next chapter will examine ways to reduce costs.

12 Reducing Costs for Ram EDM

Preparing Workpieces for Ram EDM

Since Ram EDM generally machines the entire cavity, it is sometimes cost effective to remove as much material as practical to reduce machining time for workpieces which have large cavities

Difference Between Ram and Wire EDM in Reducing Costs

There is an important difference when ram or wire EDM is used to machine parts. If a blind hex is to be ram EDMed, the hole should be drilled close to the hex as illustrated in Figure 12:1.

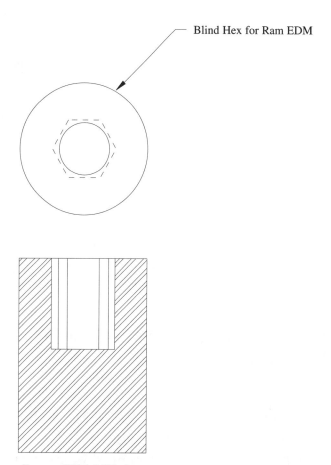

Blind Hex for Ram EDM

For ram EDM drill hole
close to the edges.

Figure 12:1
Proper Preparation for Ram EDM

If a hex goes through the workpiece and wire EDM is used, then just a starter hole should be drilled so as to make one slug. If the hole is drilled to the edge of the hex when wire EDM is used, six slugs will be produced. The wire EDM machine needs to be stopped six times to remove the fallen slugs. Machining one slug will reduce the costs significantly when wire EDM is used. See Figure 12:2.

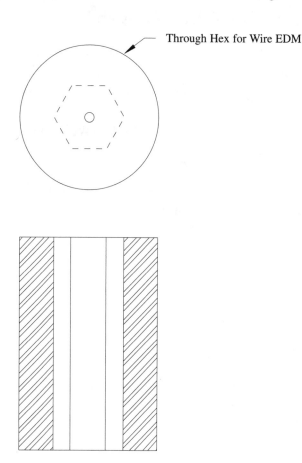

Through Hex for Wire EDM

Figure 12:2
Proper Preparation for Wire EDM
Drill only a starter hole so as to produce one slug..

Prolonging Electrode Life With No-Wear EDMing and No Premachining

Ram EDMing has the capability to cut material with relatively little electrode wear. Although cutting with no-wear settings is a slower method, it often prevents the need to premachine the cavity.

In previous years, when ram EDM was slow and electrode wear high, roughing out the cavity prior to EDMing was an established practice. Unless the cavity was premachined, costly roughing and finishing electrodes had to be made. Skilled machinists were needed to mill the pocket and to make sure the print was followed. With the advent of solid-state power supplies and premium electrode materials, it became possible to rough out a number of cavities with no-wear settings, even in hardened materials.

Certain cautions need to be applied when using no-wear settings. Premium graphite should be used. (Improper graphite can increase the wear 25%, instead of producing less than 1% wear.) Enough stock should be left for finishing because the gap between the electrode and the workpiece is much greater when roughing than when finishing.

When no-wear power setting is used for roughing, and the electrode is close to finishing, it is generally more effective to switch to negative polarity, instead of using a lower power setting in the positive polarity. Changing the no-wear cutting settings for finishing increases electrode wear and reduces cutting speed. By simply switching to negative polarity, operators can make cuts two to three times faster and produce less electrode wear. The no-wear technique is not for materials such as carbide, aluminum, titanium, hastalloy, and other high-nickel content metals.

Electrode and Workpiece Holding Devices

Various manufacturers have developed methods that greatly aid ram EDM. There are electrode holders that can be removed from the machine and reinserted into their exact locations. See Figure 12:3. This reinsert capability is especially important when worn electrodes need to be redressed

Courtesy System 3R

Figure 12:3
An Electrode Holding Kit

Electrode holders can also be used when machining the electrode. After the electrode is machined, it will be properly oriented because the same holder was used for machining and EDMing.

Palletizing workstations allow workpieces to be placed repeatedly in the required location. Rotating dividing heads allows parts to be rotated and put on an angle for machining, as in Figure 12:4.

Courtesy System 3R

Figure 12:4
Dividing Head

Orbiting

One of the most dramatic improvements in ram EDM was the introduction of orbiting. Previously, three to four electrodes were often needed to finish a cavity. A roughing electrode was first used, then two to three finishing electrodes. Unless the electrode could be recut, two or three finishing electrodes were needed because of excessive corner wear, as shown in Figure 12:5. In addition, the finishing electrodes had to be the exact dimension, minus the overcut.

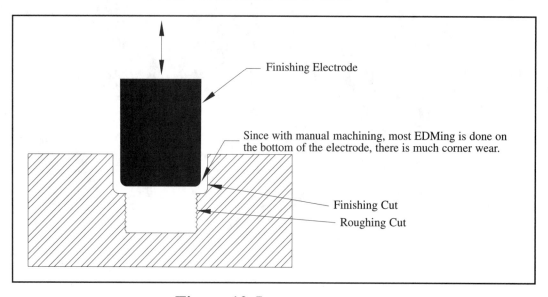

Figure 12:5
Finishing with Manual Machines

With orbiting capabilities, the roughing electrode can often be used for the finishing electrode. This dual use substantially reduces the cost for producing cavities. With an orbiting device, the exact orbit can be set so the cavity will finish to the desired dimension.

The orbital path also aids in the flushing of the cavity by creating a pumping action. Since the same electrode produces the first cavity and the finish cavity, the entire electrode is put into the cavity on the second cut. Now the electrode cuts not only on the bottom, but also along the sides of the electrode. This cutting action greatly reduces corner electrode wear, as shown in Figure 12:6.

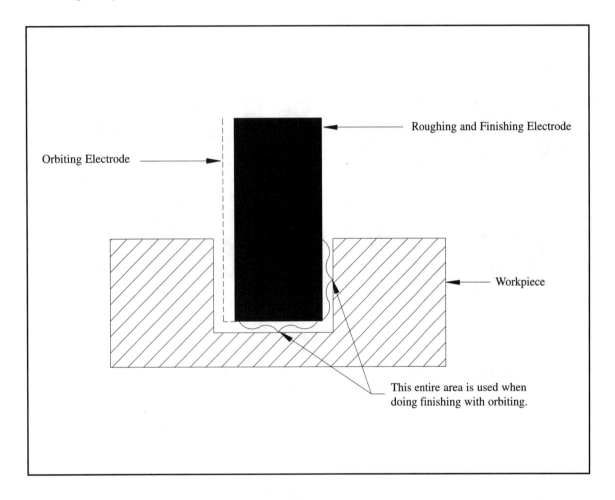

Figure 12:6
Finishing With Orbiting

Since a greater surface area is being machined when orbiting, greater current can be used. Allowing greater current settings increases cutting efficiency without sacrificing surface finish. Orbiting also decreases side wall taper.

Along with CNC came the introduction of various orbital paths, as depicted in Figure 12:7. Such orbital flexibility greatly increased the efficiency of ram EDM cutting.

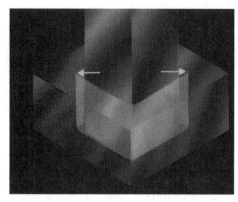

Down machining

Cycle on X, Y or Z axis is intended mainly for rough machining.

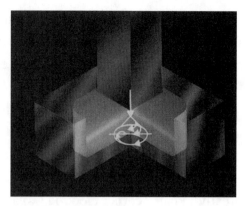

Orbital machining

Down machining followed by orbits allows machining of three-dimensional forms from roughing to finishing. Machining axis X, Y or Z.

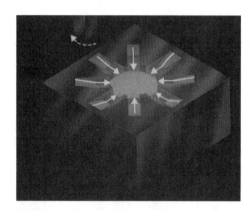

Vectorial machining

Allows cavity or form machining in any direction.

Vectorial machining

For servocontrolled machining of the electrode around its axis.

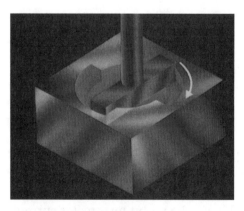

Vectorial machining

Combined with electrode rotation for machining intricate forms using simple shaped electrodes.

Directional machining

To obtain sharp corners. Machining axis X, Y or Z. The translation is automatically calculated by the CNC according to the location and the value of the angles to be machined.

Courtesy Charmilles Technologies

Figure 12:7
Various Orbital Paths

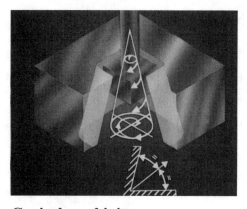

Conical machining

Of negative and positive tapers encountered, for example, in cutting tools and injection molds. Angles may be programmed from 0° to ± 90°. Machining axis X, Y or Z.

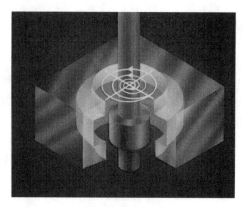

Horizontal planetary machining

For grooves, threads, etc. Machining axis X, Y or Z.

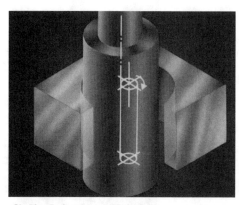

Cylindrical machining

Permits a non-servocontrolled translation movement of the electrode: for rough machining under poor flushing conditions. Machining axis X, Y or Z.

Helical machining

For threads and helical shapes.

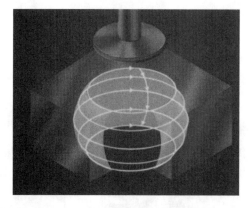

Concave spherical machining

Spherical forms can be produced using globe shaped electrodes or spherical caps with thin cylindrical electrodes. Machining axes X, Y or Z.

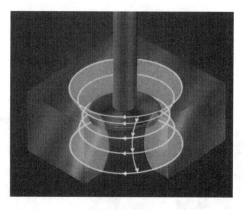

Convex spherical machining

Spherical forms can be produced using globe-shaped electrodes or spherical caps with thin cylindrical electrodes. Machining axis X, Y or Z.

Courtesy Charmilles Technologies

Figure 12:7
Various Orbital Paths

Manual Machines Mounted With Orbiting Devices

Manual machines can be equipped with orbiting capabilities. These devices are similar to a boring head on a milling machine which allows the electrode to form an orbital path. Although these manual orbiting devices are less sophisticated than CNC orbiting, they increase the cutting efficiency of the manual machines.

Repairing Molds With Microwelding

Traditionally, when nicks, scratches, worn parting lines, or other mold damages were detected, the mold was disassembled and then sent to be TIG (Tungsten Inert Gas) welded. The welder preheated the block to avoid cracking the mold and then welded the defective area. The block was allowed to return to room temperature slowly and then machined and polished. This was a time-consuming process to repair molds, even with minor repairs.

Today, microwelding units that can weld the head of a pin are available. The current discharge is of such short duration and produces such little heat that the smallest repairs can be made without damaging the surrounding area of the mold. Some repairs can be made where the mold remains in the injection molding machine.

A metal strip or wire consisting of material similar to the workpiece is placed over the area. A non-arcing spot welding process bonds the material to the workpiece. After the welding process, the applied material becomes hard. The hardness depends upon what material was used for welding. For small repairs, such as pit marks, a metal paste is used. Since the welds are not excessive, they require less machining and hand polishing. See Figure 12:8.

Courtesy Rocklin Manufacturing

Courtesy Rocklin Manufacturing

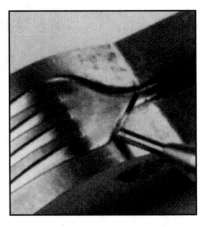

Courtesy Gesswein

Figure 12:8
Rebuilding a Worn Parting Line in a Mold with Microwelding

Abrasive Flow Machining

Some manufacturers use abrasive flow machining to remove the recast layer from EDMing. The process involves two opposing cylinders which extrude an abrasive through the desired surface. The abrasives that are forced over the EDM area polish the surface. Abrasion occurs only in the restricted area. For further information, see Chapter 14, "Abrasive Flow, Thermal Energy Deburring, and Ultrasonic Machining."

Automatic Tool Changers

For round-the-clock operation, some companies use automatic tool changers. Units are available that can carry from 50 to 100 electrodes. These robotic units can change electrodes, as well as workpieces, for unattended operations. Various automatic tool changers are also on the market. See Figure 12:9

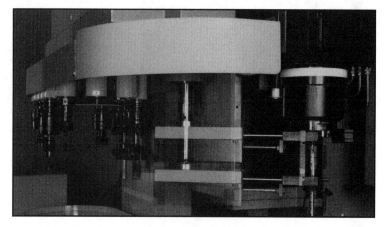

Courtesy Sodick

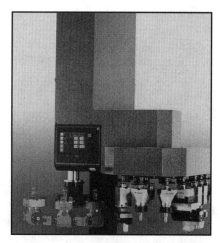

Courtesy Mitsubishi

Figure 12:9
Machines Equipped With Automatic Tool Changers

Automatic tool changers can also be added to a machine as shown in Figure 12:10.

Courtesy System 3R

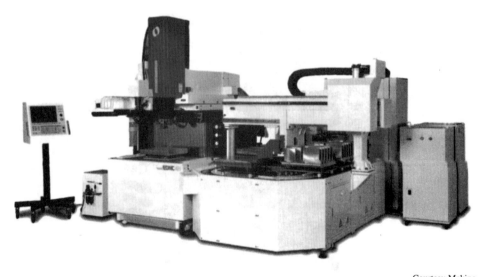

Courtesy Makino

Figure 12:10
Attaching Automatic Tool Changers

Hard Die Milling

Die molds with hardness of RC 50+ can be machined with hard die milling to being totally finished, without ram EDM. Machining molds in this manner can significantly reduce the amount of hand polishing, and in some cases altogether. However, hard die milling is more practical for open type surfaces, rather than for deep intricate cavities. See Figure 12:11.

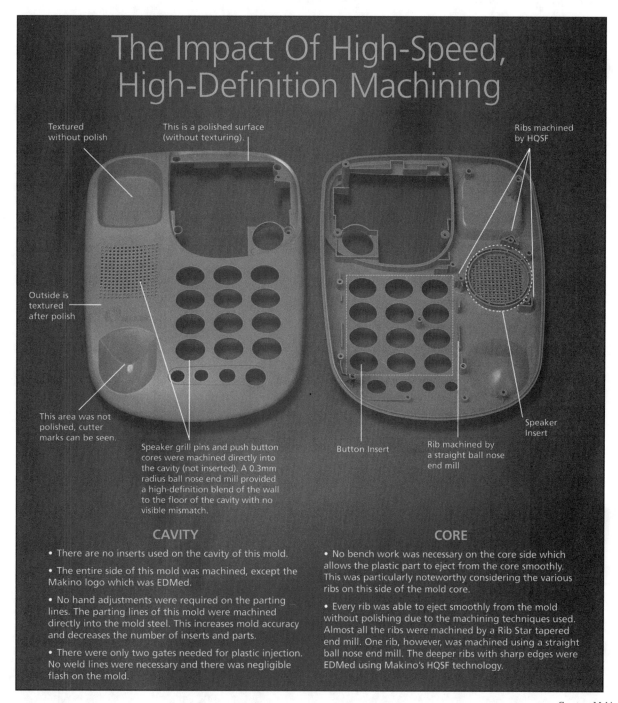

Courtesy Makino

Figure 12:11
Telephone Cavity and Core Machined Primarily With Hard Die Milling

To machine hardened materials, tools need to be tough and heat resistant. One company found that TiAIN-coated tools were best for wear resistance and material removal. It claimed that standard carbide end mills wore out too quickly.

For building a telephone mold, one manufacturer ran its high performance vertical milling center for a 12 mm (472") end mill at 6630 RPM (2590 mm/min), and a .6 mm (.024") end mill at 20,000 RPM (800mm/min). See Figure 12:12. These special milling machines are designed for high-speed milling.

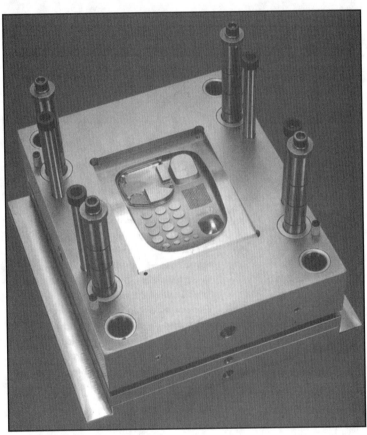

Courtesy Makino

Figure 12:12
Telephone Mold Machined With Hard Milling and Without Inserts

Future of Ram EDM

Manufactures have produced an EDM grinder and an EDM mill, but both projects have been abandoned. However, better power supplies, fuzzy logic, CNC orbiting, and robotic handling of electrodes and workpieces have increased the efficiency of ram EDM. As this process becomes better understood and utilized, it will further reduce machining costs associated with ram EDM.

Unit 4
Fast Hole EDM Drilling

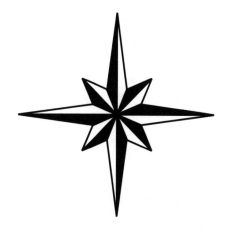

Notes

13 Fast Hole EDM Drilling

Once relegated to a "last resort" method of drilling holes, fast hole EDM (electrical discharge machining) drilling is now used for production work. Drilling speeds have been achieved of up to two inches per minute. Holes can be drilled in any electrical conductive material, whether hard or soft, including carbide. See Figure 13:1.

Courtesy Charmilles

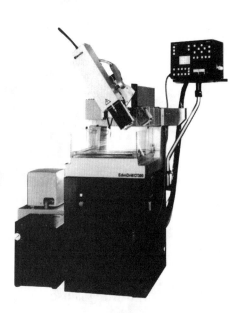
Courtesy Belmont Equipment

Courtesy Sodick

Courtesy Current EDM

Figure 13:1
Small Hole EDMs

For high-production fast hole drilling, machines are also available with tool changers. See Figure 13:2.

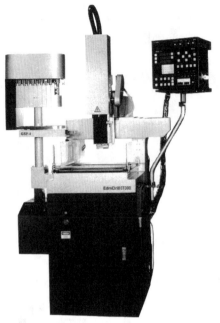

Courtesy Current EDM

Figure 13:2
Small Hole EDM with Tool Changer

Fast hole EDM drilling is used for putting holes in turbine blades, fuel injectors, cutting tool coolant holes, hardened punch ejector holes, plastic mold vent holes, wire EDM starter holes, and other operations.

The term "fast hole EDM drilling" is used because conventional ram EDM can also be used for drilling. However, ram EDM hole drilling is much slower than machines specifically designed for EDM drilling. See Figures 13:3 and 13:4.

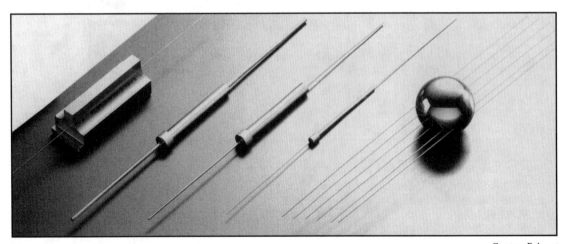

Courtesy Belmont

Figure 13:3
EDMed Drilled Parts

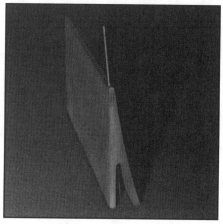

Courtesy Current EDM

Figure 13:4
Turbine Blade Drilled With EDM

How Fast Hole EDM Drilling Works

Fast hole EDM drilling, as illustrated in Figure 13:5, uses the same principles as ram EDM. A spark jumps across a gap and erodes the workpiece material. A servo drive maintains a gap between the electrode and the workpiece. If the electrode touches the workpiece, a short occurs. In such situations, the servo drive retracts the electrode. At that point the servo motor retraces its path and resumes the EDM process.

Courtesy Charmille

Figure 13:5
EDMing a Hole

A. Dielectric and Flushing Pressure

The dielectric fluid flushes the minute spherical chips eroded from the workpiece and the electrode. The dielectric fluid also provides an insulating medium between the electrode and the workpiece so that sufficient energy can be built. When the dielectric cannot resist the applied energy, a spark jumps from the electrode to the workpiece and causes the spark to erode the workpiece and the electrode. The servo mechanism provides the proper gap for spark erosion to continue.

Deionized water is used in some machines, but some manufacturers recommend an additive to aid their machines in cutting. To accomplish fast hole EDM drilling, high-pressure flushing is used (up to ten times the pressure for conventional ram EDM). Flushing pressure is one of the most important variables for high speed EDM drilling.

The dielectric should be clean. Some manufacturers use the dielectric only once; others reuse it. When the dielectric is reused, it should be filtered carefully to remove eroded particles.

B. The Electrode

A round hollow electrode is constantly rotated as the dielectric fluid is pumped through the electrode. The rotating electrode helps in producing concentricity, causing even wear, and helps in the flushing process. See Figures 13:6 and 13:7. Since the eroded particles are conductive, removing them from the hole is important to prevent shorting between the electrode and the workpiece, and to prevent EDMing the sides of the hole.

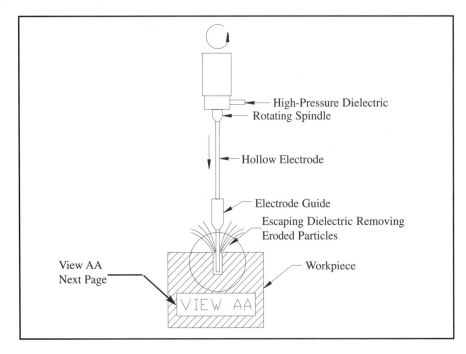

Figure 13:6
Fast Hole EDM Drilling

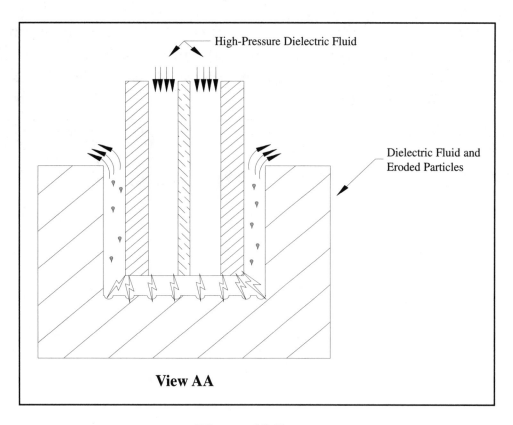

High-Pressure Dielectric Fluid

Dielectric Fluid and
Eroded Particles

View AA

Figure 13:7
Rotating Electrode Eroding the Workpiece

The high flushing pressure through the center of the electrode tends to stiffen it. Also, the dielectric being forced out of the hole produces a centering effect upon the electrode. With the aid of the electrode guide and the flushing effects on the electrode, EDM drilling can penetrate much deeper than almost any other drilling method. Holes have been drilled up to 500 times the diameter of the electrode. Using three .012″ diameter electrodes, one company drilled a .015/.016″ hole in aluminum 7 1/2″ (190 mm) deep. Another company drilled a 7″ (180 mm) hole in stainless steel within +/- .005″ (.13 mm).

Generally, deep hole EDMing is used for holes from .012″ to .125″ (.3 mm to 3.18 mm). However, .0058″ (.15 mm) holes have been drilled repeatedly with a .004″ (.1 mm) diameter electrode in .050″ (1.3 mm) thick stainless steel.

The high flushing pressure helps keep the workpiece and electrode cool. This helps to keep the heat-affected zone, or depth of recast level at a manageable level. The pressure also aids in producing a reasonably good finish. Regular ram EDM's, with weaker flushing pressures, are unable to duplicate the results of fast hole EDM machining. They are slower, the finish is not as good, and they have a higher recast layer.

Hollow electrodes allow dielectric fluid to flow through the electrode center. However, larger electrodes with a single hole can create problems. As the electrode erodes the workpiece, the center of the electrode does not remove material, thereby leaving a spike. The spike can cause the machine to short. A short causes the machine to retract, which lengthens the cutting time. To overcome this problem, electrodes with multiple channels were developed to eliminate center slugs, as shown in Figure 13:8.

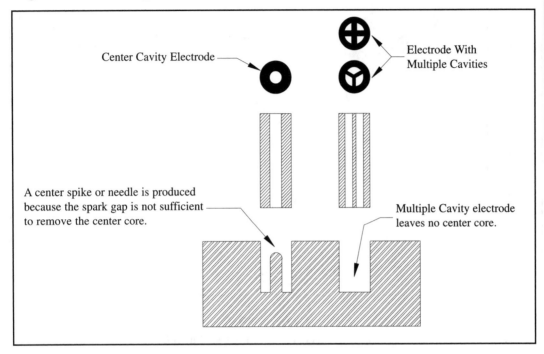

Figure 13:8
Various Tubular Electrodes and Their Results

C. Electrode Guides

The insulated electrode guide keeps the electrode on location and prevents drifting while the rotating electrode is cutting. The electrode guide prevents electrode wobbling and aids in minimizing the EDM overcut, generally .002″ (.05 mm) per side. The guides are usually about .200″ to .400″ (5 to 10 mm) above the workpiece, which allows the high pressure dielectric to escape from the hole.

D. Servo Motors

The servo motors are controlled by a microprocessor which measures the gap voltage. By monitoring the gap voltage, the servo motor maintains the proper gap for spark erosion. If the gap voltage is too high, as in a short or accumulation of debris, the microprocessor signals the servo motor to retract the electrode. When the gap voltage is reduced, the servo motor advances the electrode and resumes cutting.

Due to the high-pressure removal of the EDM chips, the servo motor needs no constant retract cycle as in conventional ram EDM. The constant forward motion allows for rapid EDMing of holes.

Metal Disintegrating Machines Compared to Fast Hole EDM Drilling

Metal disintegrating machines use the same principles as EDM, but these machines are used primarily for removing various types of broken taps, drills, and fasteners. Fast hole EDM drilling is a much more precise method for drilling.

A metal disintegrating machine uses a hollow electrode to erode broken tools or fasteners. A coolant flows through the electrode and flushes the metal particles. Since the surface finish is unimportant, these metal disintegrating machines can remove within 1 minute a broken 1/4″ (6 mm) tap that is 1″ (25 mm) in the workpiece, and within 2 minutes a 1/2″ (13 mm) tap that is 1″ (25 mm) in the workpiece. These machines also come in portable models and can cut upside down.

Other Methods to Produce Holes

Besides fast hole drilling, ram EDM, lasers, and photochemical machines can produce holes, even into hardened materials. Conventional drilling machines using carbide drills can also drill many hardened materials.

Disadvantages in Fast Hole EDM Drilling

A. Electrode Wear

Considerable electrode wear results from EDM drilling. The electrode wear can equal or exceed the depth of the hole. For example, a two inch (51 mm) depth can wear the electrode two inches (51 mm) or more.

B. Hole Quality

When EDM drilling holes quickly, then fine finishes, low recast, and heat-affected zones are sacrificed. Under ideal conditions, hole diameters and tapers per inch have been consistently drilled to +/- .0003″ (.008 mm) in 1″ (25.4 mm) thick material. In general, these conditions are not achieved in fast hole drilling. For many applications, however, the process is more than satisfactory.

C. Inability for EDMing Miniature Holes Below .012″ (.3 mm)

The necessity of having a hole in the electrode makes drilling miniature holes less than .012″ (.3 mm) very impractical. Manufacturing these hollow miniature electrodes is difficult. Furthermore, the thinner electrodes are not rigid enough to carry the high-pressure flushing without deflecting. Also, wear on small electrodes can be from ten times the hole depth.

D. Reduced Speed for Large Holes

Although large holes can be EDMed, the drilling time is often not competitive with conventional drilling or with wire EDM. For some difficult drilling applications, like carbide, a starter hole can be drilled with fast hole EDM and then machined with wire EDM. Fast hole EDMing is also used for holes that cannot be deburred due to obstructions.

E. Blind Holes are Difficult to Control

Due to the high electrode wear, the depth of blind holes is difficult to control. Whenever possible, conventional drilling should be used for blind holes.

However, if a blind hole is needed, the electrode needs to be dressed. Otherwise, electrode wear causes a bullet-shaped hole at the bottom. This bullet shape needs to be removed to obtain a square-edged bottom hole.

Advantages in EDM Drilling

A. Drilling on Curved and Angled Surfaces

When holes must be drilled on curved or angled surfaces, great difficulties arise with conventional drilling. Drills tend to walk off such surfaces. To prevent drills from walking, fixturing and guide bushings are used on these irregular surfaces to guide conventional drills. But in EDM drilling, the electrode never contacts the material being cut. This non-contact machining process eliminates the tool pressure when drilling on curved or angled surfaces. See Figure 13:9.

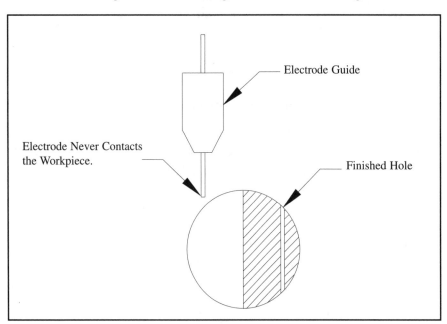

Figure 13:9
Non-contact machining allows electrode to enter curved and angled surfaces.

B. No Hole Deburring

Deburring of holes from conventional drilling can take longer than drilling the holes. As in conventional EDMing, fast hole EDM drilling creates no burrs when drilling. See Figure 13:10. This burr-free drilling is especially important when difficult holes, such as turbine blades, require deburring.

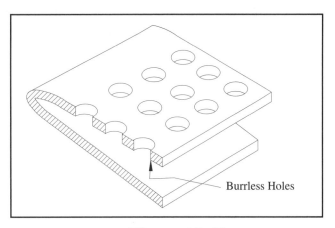

Figure 13:10
Difficult to Deburr Holes

C. Drilling Hardened Materials

Some materials are too hard to drill using conventional methods, i.e., hardened tool steel, difficult alloys, and carbide. But material hardness does not affect the EDM process. However, some materials, like carbide, cut slower, not because of hardness, but because of conductivity properties of carbide.

D. Materials That Produce Gummy Chips

Materials such as soft aluminum and copper can produce gummy chips. EDM drilling easily machines such materials.

E. Drilling Deep Holes

Drilling deep small holes with conventional drilling is often extremely difficult, and many times impossible. Fast EDM hole drilling is often the only practical method for producing such holes.

F. Preventing Broken Drills

As conventional drills enter or exit curved or angled surfaces, they tend to break if not carefully controlled. Small broken drills are also often extremely difficult to remove from the workpiece. To prevent breaking drills in conventional drilling, controlling torque conditions are critical. However, in EDM drilling the torque conditions do not exist since the electrode never contacts the workpiece.

G. Creating Straight Holes

Due to the non-contact process of EDM, the deep hole EDM drilling produces straight holes. In contrast, conventional deep hole drills tend to drift.

Accuracy of Fast Hole EDM Drilling

Because eroded particles from the holes are flushed, variations occur in the hole diameter. These are the results of fast hole EDM drilling with a .040″ (1 mm) drill in D2 tool steel:

Depth Straightness	Taper
1″ (25.4 mm) +/-.0003″ (.0076 mm)	+/-.0005-.001″ (.013-.025 mm)
4″ (102 mm) +/-.001-.0015″ (.025-.038mm)	+/-.0025-.004″ (.064-.102 mm)
8″ (203 mm) +/- .0015-.004″ (.038-.102 mm)	+/-.005″ (.127 mm)

Versatility of Fast Hole EDM Drilling

At Reliable EDM, we purchased a fast hole EDM drilling unit that could be mounted on a milling machine to obtain greater versatility. This enabled us to EDM large workpieces. See Figure 13:11.

Figure 13:11
Fast Hole EDM Drill Mounted on a Milling Machine

Conclusion

Fast hole EDM drilling has many applications. It is an extremely cost effective method for producing fast and accurate holes into all sorts of conductive metals, whether hard or soft.

Unit 5

Abrasive Flow, Thermal Energy Deburring, and Ultrasonic Machining

Notes

14 Abrasive Flow, Thermal Energy Deburring, and Ultrasonic Machining

Abrasive Flow Machining

Abrasive flow machining (AFM) uses a semisolid abrasive media to deburr, polish, remove recast layers, or put radii on sharp corners. Some systems are manual and others are automated with robots. See Figures 14:1 and 14:2.

Courtesy Dynetics Corporation

Figure 14:1
An Abrasive Flow Machine

Courtesy Extrudehone

Figure 14:2
A Fully Automated Abrasive Flow Machine

Various parts have been processed using this method of machining.

Courtesy Extrudehone

Figure 14:3
Parts Processed With Abrasive Flow Machining

A. How Abrasive Flow Machining Works

Abrasive flow machines force an abrasive media through or over unfinished surfaces. One system uses two opposing reciprocal cylinders that force abrasive media over the surfaces. When one cylinder finishes pushing the media through the workpiece into the receiving cylinder, the receiving cylinder reverses the process. With this process both openings of the workpiece need to be sealed to contain the pressurized media.

Another means of abrasive flow machining forces abrasive media one way through the workpiece as shown in Figure 14:4. This process uses only one cylinder to force the media through the workpiece. After the cycle is finished, compressed air is used to clean the part of the remaining media. The advantage of this abrasive flow machining is it needs less tooling to seal the media inlet, and permits easier workpiece loading and simpler workpiece media removal.

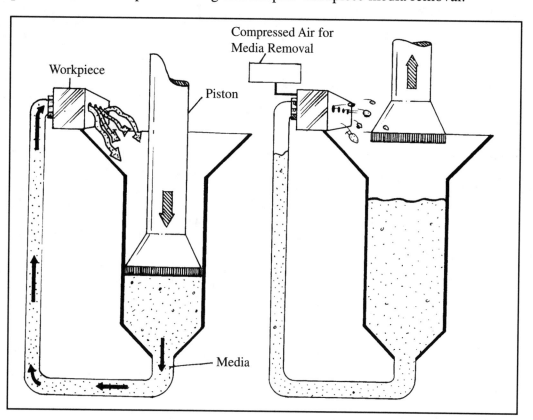

Courtesy Extrudehone

Figure 14:4
Abrasive Flow Machining

To insure proper abrasive flow machining, the media flow area needs to be sealed. Only where the media is restricted will abrasive machining occur. As the compressed media flows over the surface, it polishes or deburrs the contact area as shown in Figure 14:5.

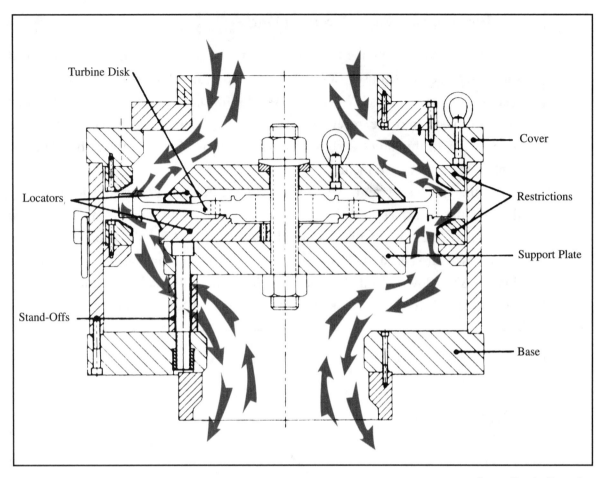

Turbine Disk

Locators

Stand-Offs

Cover

Restrictions

Support Plate

Base

Figure 14:5
Abrasive flow fixture for radiusing the edges of broached
slots on turbine and compressor disks.

There is a basic rule for abrasive flow machining. The abrasive compounds act most aggressively where the compound achieves greatest velocity. If a part has small holes and the media is forced through them, the media will deburr, radius, and polish these small holes while leaving large feed holes and other surfaces unaffected. This selective action occurs because the media velocity is greatest going through the small holes.

Using abrasive flow machining for its exhaust manifolds, one automotive manufacturer increased the horsepower of its engines by four horsepower, and at the same time reduced emissions and increased fuel mileage. Processing time was less than two minutes.

Sand-cast aluminum passages of the intake manifold measured a surface roughness of over 200μ inch R_a. Using abrasive flow machining (Figure 14:6), the surface finish was lowered to under 20μ inch R_a. The abrasive flow machining process reduced flow variation between cylinders, resulting in a more efficient engine and less exhaust emissions.

Courtesy Extrudehone

Figure 14:6
Abrasive Flow Machining an Intake Manifold

B. Abrasives

The abrasives generally used for abrasive flow machining are aluminum oxide, silicon carbide, boron carbide, and diamonds. Silicon carbide performs well for most applications. The high costs of boron carbide and diamond limit their use to hard materials.

Abrasive flow machining process is similar to lapping. Sharp media grains rub and gradually polish the surface. Abrasive grit size determines the speed of abrasion and surface finish. The smaller the grit, the slower the process—but the finer the finish. Abrasive grit sizes range from a fine powder, 1000 grit, to 10 grit and larger.

C. Viscosity

Along with the abrasive type, the viscosity of the media greatly affects the machining rate and finish. A stiff media (high viscosity) acts more like an extrusion process and is used primarily for polishing. A thinner media (low viscosity) creates greater radii at the passage openings, and is used as a deburring media. See Figure 14:7.

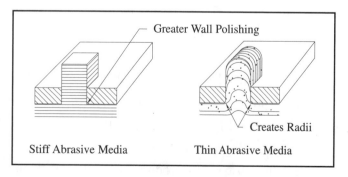

Figure 14:7
Effects of Media Viscosity

The abrasive media forces its way through the workpiece to become a formable grinding stone. Its stiffness is determined by the media viscosity. See Figure 14:8.

Courtesy Extrudehone

Figure 14:8
Abrasive Media Being Extruded

D. Extrusion Pressure

The pressure used to extrude the abrasive media has a significant effect on the machining. The higher the pressure, the greater the machining results. The amount of the extrusion pressure, the strength of the workpiece, and the viscosity of the media all affect the amount of pressure that can be applied.

E. Finishes

Proper abrasives and viscosities can produce very fine finishes. Carbide compacting dies, with an original finish of 80μ inch R_a, were reduced to 8μ inch R_a. Finishes below 3μ inch are common. The roughness from cast, machined, or EDM surfaces can be reduced from 75 to 90 percent.

Abrasive flow machining results in fine finishes for complex surfaces and contours of impellers for aircraft engines, high-performance compressors, and automotive turbo chargers. See Figures 14:9 and 14:10. Over half of the world's makers of precision extrusion dies use this process for polishing their dies.

A manufacturer used this process to produce a better finish on a diesel truck turbocharger impeller. The cast finish on the impeller produced a surface finish of

Courtesy Extrudehone

Figure 14:9
Extrusion Dies

Courtesy Extrudehone

Figure 14:10
Abrasive flow machining is particularly suited
for surface finishing of complex surfaces.

150μ inch R_a. After using abrasive flow machining, the surface finishes were reduced to under 16μ inch R_a.

Aircraft turbine engine manufacturers use abrasive flow machining for finishing complex contours on high-precision turbine engine parts. Hand polishing time was reduced by over 85 percent.

Thermal Energy Deburring

Thermal energy deburring uses heat to deburr parts. Parts are placed inside a sealed container (Figure 14:11), which is then pressurized with a combustible gas and oxygen. A spark plug ignites the mixture, creating extreme heat inside the chamber. Excess oxygen in the chamber combines with the thin burrs and turns them into oxide powder.

Courtesy Surftran

Figure 14:11
Thermal Energy Deburrs Bulk Parts

The heat is of such short duration that it does not significantly change the material mass dimensionally or metallurgically. This process is effective even with external, internal, or blind holes. See Figure 14:12 for various parts deburred with thermal energy.

STAINLESS STEEL FITTING
Burrs removed, without the use of media
which could lodge in internal passages.

STEEL INJECTOR PARTS
1,440 pph deburred with zero defects in
one operation.

STEEL SCREW MACHINE PARTS
Bulk loaded. Burrs removed from O.D.
threads and I.D. bore.

ZINC LOCK CYLINDERS
Flash and thin webs removed during bulk
load processing.

Courtesy Surftran

Figure 14:12
Part Deburred With Thermal Energy

Disadvantages and Advantages of Thermal Energy Deburring

The limitations of thermal energy deburring are:

- Parts must be able to fit into a chamber.
- Thin or fragile sections cannot be processed.
- Some materials have such high oxidation resistance that this process is ineffective.

The advantages of this process are:

- It is more reliable than hand deburring because it removes all burrs regardless of location.
- It can be used with different metals and thermoplastics.
- Parts can be batched together.

Ultrasonic Machining and Polishing

Machining of detailed and complex geometric shapes is time-consuming and difficult. Ultrasonic machining greatly speeds this process. It is particularly useful for making graphite electrical discharge machining electrodes containing intricately detailed configurations. This process can also be used for polishing and machining glass, carbide, and ceramics.

How Ultrasonic Machining Works

A metal-forming tool, called a "sonotrode," is fabricated to the desired pattern. The sonotrode is mounted to the ultrasonic machine and over it is placed a graphite electrode.

As the electrode is lowered into the sonotrode, nozzles eject a water-based abrasive slurry between the electrode and the sonotrode. The sonotrode vibrates with a frequency of about 20,000 cycles per second, causing the fine abrasive particles to vibrate in the gap and erode the graphite electrode as shown in Figure 14:13. The form tool does not abrade the electrode. Excited abrasive particles of the abrasive slurry (usually the abrasives of silicon or boron carbide mixed with water) vibrate against the electrode and perform the machining. This process produces a mirror image of the form tool. The particle size of the grit determines the workpiece overcut, usually twice the grit diameter.

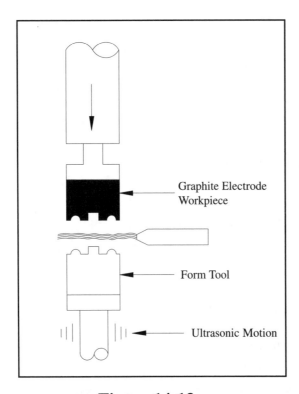

Figure 14:13
Ultrasonic Machining

Since ultrasonic machining has low machining forces, delicate electrodes can be manufactured. Details and radii of up to .002 inch (.050 mm) can be maintained with extremely fine finishes of 8 to 16µ inch. Accuracies are possible to +/- .0004″ (.010 mm). See Figure 14:14.

Courtesy Sonex, a Division of Extrudehone

Figure 14:14
Ultrasonic Machined Electrodes

The abrading process removes 1.5 to 3.5 cubic inches (25 to 60 cc) per hour and can penetrate up to 2 inches (51 mm). Because this process causes little wear on the sonotrode, many graphite electrodes can be produced from one sonotrode. Instead of producing multiple electrodes, worn electrodes can be quickly redressed with this process. The redressing operation usually takes from 2 to 10 minutes. Ultrasonic machining is especially useful for manufacturing electrodes for coining, embossing, and injection molding dies This process is also used to polish molds and die cavities. See chapter 10, "Ram EDM Electrodes and Finishing," for further discussion on abrading large graphite electrodes.

Conclusion

Abrasive flow machining can speed the process of finishing, deburring, or putting small radii on workpiece cavities. In addition, it can also finish difficult areas, such as fuel spray nozzles.

Ultrasonic machining greatly increases the speed of producing complex electrodes and other materials. It can also save countless hours in finishing molds and die cavities. For the right applications, abrasive flow and ultrasonic machining have great potential for reducing manufacturers' machining and finishing costs.

Unit 6

Photochemical Machining

Notes

15 Photochemical Machining

Photochemical machining (PCM), also known as photochemical milling, photo-etching, and photolithography, is a process that blanks or etches out parts by means of chemicals, as opposed to using abrasives or hard tooling. See Figure 15:1. Protected areas are masked, and chemicals are used to dissolve the unmasked areas. The photochemical machining process has been in commercial operation since the mid 1950s. It can process either sheets or continuous coils.

Courtesy Buckbee-Mears

Figure 15:1
Parts Produced With Photochemical Machining

Fundamentals of Photochemical Machining

Designing the Part

Normally, a computer-aided design (CAD) is made for the desired shape, and an etch line is put around the shape in order to remove the part from a metal sheet. The usual width of this line is twice the metal thickness. Often small tabs are used to hold the part during processing. After the design is completed, it is then plotted.

Imaging

The actual size of the image is then projected with a precision process camera onto film or a glass master. Glass masters are used for precision work because they are dimensionally stable and flat. Multiple negatives of the same image can be made with a process camera, or they can be reproduced on a photorepeating machine from the film or glass master. To create high-precision images, a process using laser-controlled photo plotters is used.

The metal surface to be etched is first cleansed to receive a photosensitive resist coating. Figure 15:2 shows coil coating of liquid resist for high volume production.

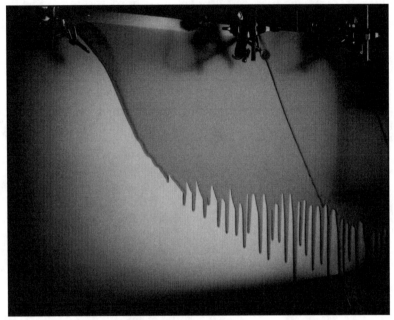

Courtesy Buckbee-Mears

Figure 15:2
Coil Coating of Liquid Photo Resist for High Volume Production

The negative images from the film or glass masters are then exposed on the photo resist-coated metal and developed. The image may be produced on either one side or both sides of the coated metal. See Figure 15:3 for a developer system.

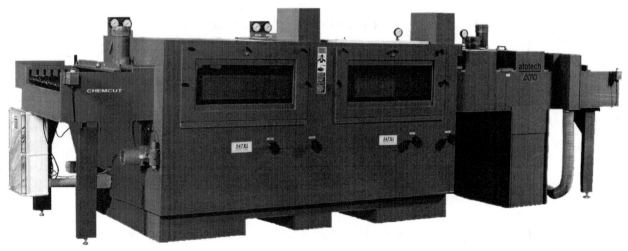

Courtesy Atotech

Figure 15:3
A Developer System

Etching and Stripping

The metal is chemically processed using a corrosive solution that is sprayed under pressure onto the photo resist coated metal. The constant force of the spray washes away the reaction products. Where the photo resist has been exposed, the metal is etched away. See illustrations Figures 15:4 and 15:5

Courtesy Atotech

Figure 15:4
Spray System Washing Away Exposed Areas

Courtesy Atotech

Figure 15:5
An Etching System

Depending on the duration of the chemical process, the etching can cut entirely through the metal or just scribe lines on the metal surfaces. After parts have been chemically etched, they undergo a stripping process where the photo resist is stripped off the metal surface. See Figure 15:6 for the photochemical machining process.

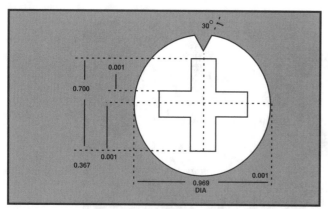

Part is drawn with CAD.

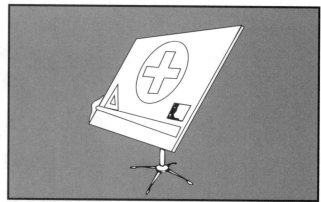

Enlarged master drawing is secured with tabs.

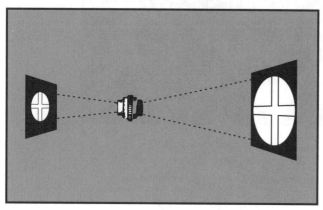

Image is reduced to exact size and exposed on film or on a glass master.

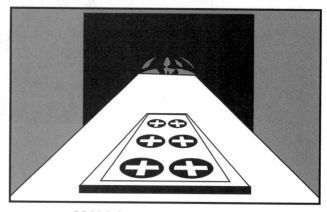

Multiple images are exposed to a metal coated with photo resist.

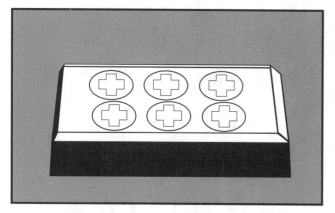

A chemically corrosive solution is sprayed on the metal. All unprotected areas will be etched away.

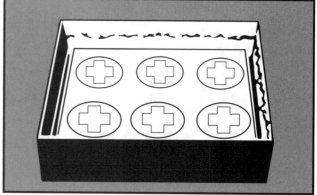

Part goes through a stripping process to remove chemicals.

Courtesy Photo Chemical Machining Institute

Figure 15:6
The Photochemical Machining Process

Materials and Products for Photochemical Machining

Many metals can be chemically etched, such as: brass, copper, nickel, silver, stainless steel, carbon steel, bronze, aluminum, tungsten, molybdenum, titanium, and zirconium. Some metals have been chemically machined from .0005 to .062″ (.013 to 1.57 mm). Copper and aluminum alloys can be machined up to .100″ (2.54 mm).

Other materials that have been photochemically machined are: plastics, polyesters, polymides, epoxy resins, glass, and ceramics. Various etchant chemicals and photoresists are used to make this process possible on such materials.

The printed circuit industry is a major user of this process. Other industries use this process in producing parts, such as: motor laminates, encoder discs, flapper valves, shims, contacts, screens, springs, heat sinks, decorative plaques, and ornaments.

Tolerances

A number of variables can affect the tolerances of parts: sheet and part size, type of metal and its thickness, and processing method. For general purposes, a tolerance of +/- 20% of the metal thickness can be used. Tighter tolerances can be obtained for certain designs. The generally stated relationship of the hole diameter to metal thickness is that the hole cannot be less than the metal thickness.

Corner Radii

The photochemical machining process tends to round corners. However, round corners are often preferred on parts. If sharp corners are required, then serifs can be added to the photo image so that when the part is etched, it will produce a sharp corner. See Figure 15:7.

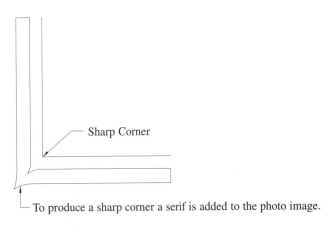

Sharp Corner

To produce a sharp corner a serif is added to the photo image.

Figure 15:7
Producing Sharp Corners

Beveling

One of the important considerations of photochemical machining is that this etching process causes bevels. The bevels can be etched so that they are equal, they are only on one side, or their depths are varied. Varied bevel depths are referred to by the percentage of etch from each side of the material, i.e. 80/20. See Figure 15:8.

Equal Beveling   One Side Beveling   80/20 Beveling

Figure 15:8
Beveling

When the metal is being etched, it is etched both vertically and horizontally. This etching process goes under the photo resist and causes bevels to occur on the walls of the material, as shown in Figure 15:9.

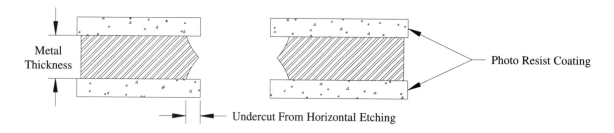

Figure 15:9
Vertical and Horizontal Etching

The depth of the bevel is determined by the thickness of the metal being etched. The thicker the material, the larger the undercut. Generally, etching on both sides of the metal creates a bevel equal to the metal thickness (1t). Etching on one side, generally produces a bevel four times the metal thickness (4t). See Figure 15:10.

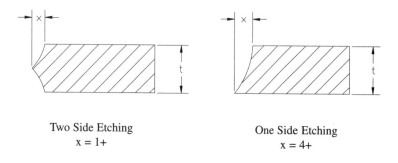

Two Side Etching   One Side Etching
x = 1+   x = 4+

Figure 15:10
Depth of Beveling

Cutting and Etching in One Operation

Parts can be cut and etched in a single operation. To accomplish cutting and etching at the same time, the film is exposed on both sides of the metal for cutting, and the film is exposed on only one side for etching, as shown in Figure 15:11. See Figure 15:12 for decorative edged products.

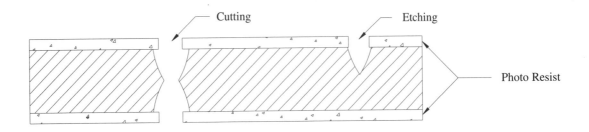

Figure 15:11
Cutting and Etching in One Operation

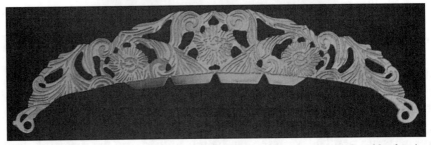

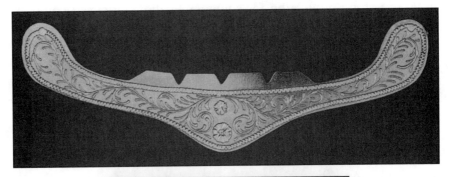

Courtesy Grace Manufacturing

Figure 15:12
Decorative Etched Products

Three Dimensional Chemical Machining

Multi-layer photo tooling makes it possible to do three-dimensional chemical machining. The partially etched stock is stripped and reprocessed between exposures. By photo etching on both sides, and masking various areas on either side, one can produce a cutout and a three dimensional shape.

Disadvantages of Photochemical Machining

A. Bevel Slots and Holes

Photochemical machining always leaves a beveled edge, but sometimes parts require straight edges. For parts with multiple holes or slots, lasers are a good machining substitute. For thin parts without many holes, stacked sheets can be cut profitably with wire EDM, which also provides a perfect, burr free straight edge.

B. High Run Production

When there is high production for some applications, it can become more cost effective to produce hard tooling.

C. Limited Metal Thicknesses

Although photochemical machining can cut copper and aluminum up to .100″ (2.54 mm), it is generally used to cut thin materials. For thick materials, other processes are used.

Advantages of Photochemical Machining

A. Eliminates the Need for Hard Tooling

Hard tooling is expensive, especially when there are many intricate shapes and holes.

B. Just In Time Machining

Rapid turnarounds are possible since no hard tooling is required. The set up and production can be accomplished in one day.

C. Freedom of Burrs

Photochemical machining produces no burrs.

D. Stress-Free Machining

When parts are blanked, stresses can be introduced into the material. Since there are no forces being applied with chemical etching, the metal remains flat. This is particularly important when machining thin pieces. Also, work hardening does not occur with this process.

E. Delicate and Complex Parts Can be Produced

The complexity of the part does not affect the cost or the processing time. Practically any drawn shape can be machined. With other processes, such as lasers and wire EDM, the length of cut is an important cost factor, but not with photochemical machining. The film is made with one exposure, and the etching process is done at one time.

Extremely complex and delicate parts can be produced with very narrow spaces because of the stress-free nature of the machining process.

Photochemical Machining of Micro-Etched Screens

Very fine screens can be produced with photochemical machining, as shown in Figure 15:13. As a general rule, hole diameters cannot be less than the material thickness. For example, a .002″ (.050 mm) thick sheet can have .003″(.076 mm) diameter holes. A .010″ (.25 mm) thick sheet can have a .012″ (.30 mm) diameter holes.

Courtesy Buckbee-Mears

Figure 15:13
Miscellaneous Micro-Etched Metal Screens

The minimum land width between holes for material under .005″ (.128 mm) is .002″ (.050 mm). For material from .005 to .010″ (.128 to .25 mm), the minimum land width is .003″ (.067 mm).

As can be observed by these figures, extremely fine screens can be made. A

90/10 screen can be produced where the bevel is larger on one side than the other. This helps greatly when a screen has to be back flushed. Normal woven wire screens have many crevices where particles can be lodged. With photochemical machining there are no crevices for particle entrapment or bacterial contamination. See Figures 15:14.

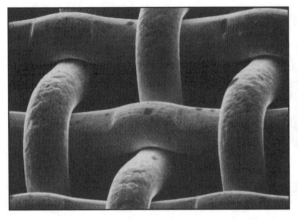 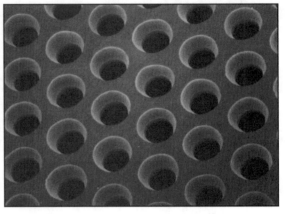

Courtesy Buckbee-Mears   Courtesy Buckbee-Mears

Figure 15:14
Wire and Photochemical Machined Screens

Electroformed Process for Micro-Etched Screens

Another process exists that is the reverse of etching, chemical deposition. It deposits metal electrically on a pattern substrate. With this electroformed process, the entire meshed surfaces are rounded, as pictured in Figure 15:15.

Courtesy Buckbee-Mears

Figure 15:15
An Electroformed Screen

Conclusion

For many applications, photochemical machining is an ideal and a cost effective method to machine parts. The ability to machine extremely complicated and detailed parts at one time makes this process a unique machining method and useful for many applications.

Unit 7

Electrochemical Machining

Notes

16 Electrochemical Machining

Electrochemical machining (ECM) is one of the least used non-traditional machining methods in spite of its high metal removal rate and relatively no electrode wear. This process is used for automotive components, gun barrel rifling, steam turbines, high production machining operations, and for deburring. See Figure 16:1.

Courtesy Extrude Hone

Figure 16:1
Airfoil Contours Produced with Electrochemical Machining

How Electrochemical Machining Works

Electrochemical machining is similar to ram EDM, since it uses a shaped electrode (the cathode), to machine the workpiece (the anode). With ram EDM, the workpiece is submerged in a tank of dielectric fluid as electrical discharges from an electrode erode material from the workpiece. With electrochemical machining, a pressurized conductive salt solution, the electrolyte, flows around the workpiece as an electrical-chemical reaction deplates material from the workpiece. See Figure 16:2 for the electrochemical process.

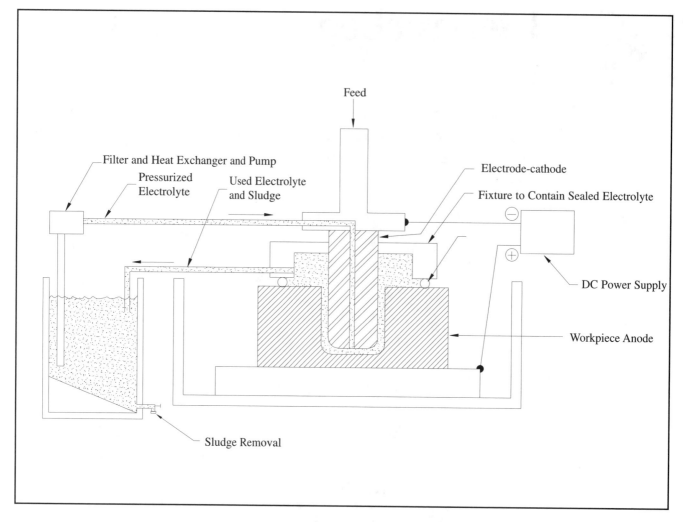

Figure 16:2
The Electro-Chemical Process

As the electrolyte flows between the narrow gap of the conductive workpiece and the shaped electrode, an electrical current of low DC voltage and high amperage is applied. The current is insufficient to produce a spark; instead, the metal dissolves from the workpiece by electrochemical reaction without any noticeable electrode wear. The controlled deplating occurs as electrical current is applied and the metal ion of the workpiece removes to an ion in the electrolyte solution.

The process of electrochemical machining is the opposite of plating. With plating, metal is applied with the use of electrical current; with electrochemical machining, metal is removed with the use of electrical current.

The electrolyte, pumped at high velocity between the electrode and the workpiece, removes the dissolved metal and heat. The electrolyte is then pumped into a tank where the sludge is eliminated, and a heat exchanger removes the heat. The filtered cooled electrolyte is then reused. See Figure 16:3 for an electrochemical machine..

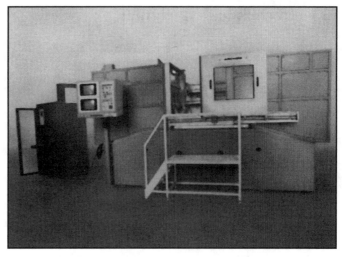

Figure 16:3
An Electrochemical Machine

Disadvantages of Electrochemical Machining

A. The Shaped Workpiece is Not a Replica of the Electrode

The flow of the electrolyte and the shape of the electrode are critical to the shape that will be made in the workpiece. In other words, a square electrode will not necessarily produce a square form with electrochemical machining. The electrolyte flowing around the electrode determines the produced shape. With ram EDM, the shape produced in the workpiece is a replica of the electrode.

B. Shaped Electrodes are Difficult to Machine

The electrochemical machining process requires skilled craftspersons to produce electrodes, because the electrodes are not replicas of the workpieces. This process is generally used only for moderate to high production items.

C. Electrolyte and Sludge Removal

The highly corrosive electrolyte is a salt solution with additives, also the electrochemical machining process produces metal hydroxides, or hydrates, that become sludge. Filters, settling, or centrifugal devices are used to remove the sludge from the electrolyte. The amount of sludge removal can be 100 to 500 times the volume of the material removed. The sludge can be put into a filter press where it is condensed to 40 to 50 percent solid matter, or the metal can be removed from the sludge. This sludge removal and treatment process is an important factor when considering electrochemical machining. Because sludge can be hazardous to the environment, one large electrochemical machining company removes all the metal from the sludge. This company produces from four to five tons of metal a week from the sludge.

Advantages of Electrochemical Machining

A. Practically No Electrode Wear

A great advantage of the electrochemical machining process it produces practically no wear on the electrode. Thousands of parts can be made with just one electrode.

B. No Recast Layer or Thermal Stress

With EDM, there is a recast layer; however, with electrochemical machining there is no recast. EDM uses spark erosion to remove metal. Electrochemical machining dissolves the metal and produces no thermal stress.

C. Material Hardness and Toughness Not a Factor

Electrochemical machining removes metal atom by atom; the material hardness and toughness has little effect on this process.

D. Rapid Metal Removal

Electrochemical machining rapidly removes metal. A large electrochemical machining job shop applies this rule: with 10,000 amps of power—one cubic inch of metal removal per minute. This particular job shop reports that the jet engine business provides 99% of its work. Although electrodes are difficult to produce, the company says that it produces jet engine air foils holding +/- .002″ (.051 mm).

E. Deburring and Radiusing of Holes

An ideal use of ECM is the deburring and radiusing of holes as illustrated in Figure 16:4.

Courtesy Dynetics

Figure 16:4
Various Parts That Have Been Deburred With Electrochemical Machining

Figure 16:5 shows what happens when the electrode in electrochemical machining remains stationary in relation to the workpiece. The current flows between the electrode and the workpiece, but the machining takes place only on the workpiece. This controlled machining results in deburred or radiused holes.

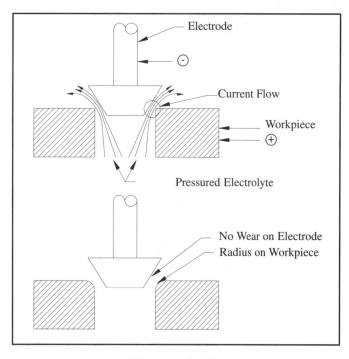

Figure 16:5
Electrochemical Deburring and Radiusing

One company uses the electrochemical machining process to deburr and radius holes for automotive air bag inflator housings. Typically, these housings contain fifty holes that allow gas to inflate the air bags within milliseconds. The hole edges are critical to the performance of these units. ECM can deburr and put radiuses on eight parts in less than ten seconds. The fully automated machine loads, unloads, and washes the air bag inflator housings—it produces 20,000 parts a day. See Figure 16:6

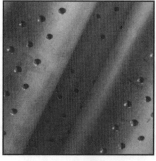

Courtesy Extrude Hone

Figure 16:6
Deburring and Radiusing of Holes

Stem Drilling

The stem drilling process drills small deep holes. Electrolyte is pumped through a hollow acid resistant tube, usually made of titanium. Except for the tip, the entire tube is insulated. As the tip approaches the workpiece and the electrolyte flows from its center, the workpiece erodes. The used electrolyte flows past the insulated tube and the walls of the hole. Often the electrolyte is filtered and reused. Various sized and shaped electrodes can be used at the same time. See Figures 16:7 and 16:8

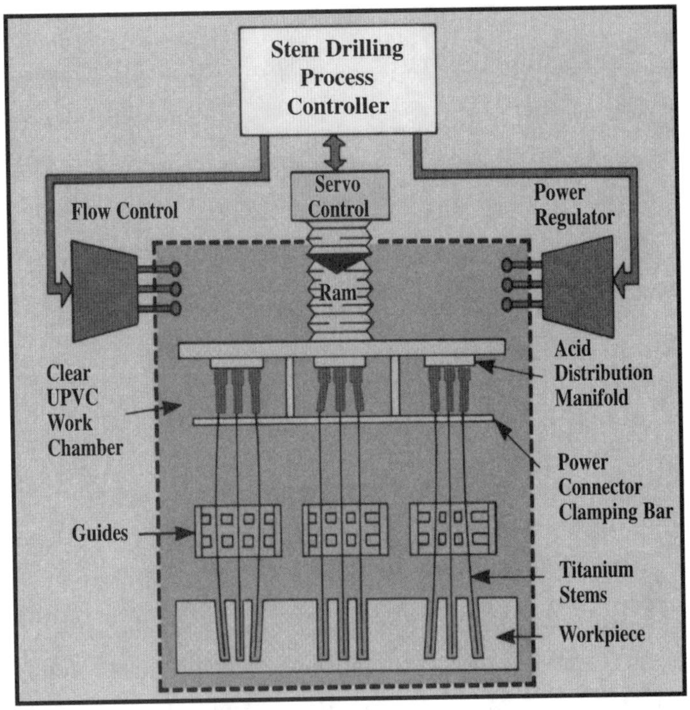

Courtesy Raycon/AMCHEM

Figure 16:7
Multi-Part Stem Drilling System

Courtesy Raycon/AMCHEM

Figure 16:8
Gas Turbine Vane With 54 Trailing Edge Holes Drilled In One Operation

Capillary Drilling

Capillary drilling is similar to stem drilling except that the tube consists of glass with a metal wire running through the tube center. As the electrolyte flows through the fine glass tube and electricity flows through the wire, the workpiece erodes at the tip. This process is used when the hole diameters are under .016″ (.4 mm) and the depths exceeds the ratio of 10 to 1.

Conclusion

Few companies use electrochemical machining, in spite of its advantages of high metal removal rate, no electrode wear, and ability to cut difficult metals. The major difficulties for electrochemical machining are:

1. The highly corrosive effects of the electrolyte.
2. The electrode not being a replica of the workpiece.
3. The difficulties in producing the electrodes and the tooling.

In spite of electrochemical machining drawbacks, there are useful applications for this process. Electrochemical machining can be profitable for parts requiring high surface quality, high production, and when other machining methods are inefficient.

Unit 8

Plasma and Precision Plasma Cutting

Notes

17 Plasma and Precision Plasma Cutting

Plasma has been used for thermal cutting for many years. The heat created by plasma cutting machines is nine times hotter than the surface of the sun. The light from stars results from plasma gases.

Plasma cutting systems use a pressurized conductive gas to cut electrically conductive materials. The difference between regular plasma and precision plasma, also call high-definition plasma and fine plasma cutting systems, is that the precision system cuts more accurately due to increased energy density. See Figures 17:1 and 17:2.

Courtesy Hypertherm

Figure 17:1
Conventional Plasma

Figure 17:2
Precision Plasma

How Plasma Cutting Works

The common concept of matter is that it has three states: solid, liquid, and gas. But there is a fourth state: plasma. For example: the element water can be ice, liquid, steam, or plasma. The amount of heat energy applied to each state changes its qualities.

Adding energy to ice results in water. Adding more energy to water creates steam consisting of hydrogen and oxygen. Adding additional energy to steam produces ionized gas, which becomes electrically conductive. This electrically conductive ionized gas is called a plasma. A lightning bolt is an example of plasma. See Figure 17:3.

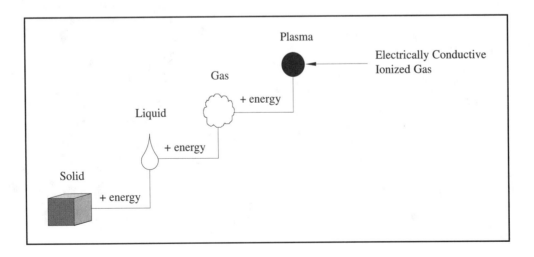

Figure 17:3
The Four States of Matter

The interior of a plasma cutting torch consists of a consumable nozzle, an electrode, and either gas or water to provide cooling. When the gas flow stabilizes, the power supply provides a high-frequency current to the electrode. This current creates electric arcs inside the nozzle. The flowing gas passes through these arcs, becomes ionized, and results in an electrically conductive plasma arc. Pressurized gas flowing through the nozzle then creates a pilot arc as shown in Figure 17:4.

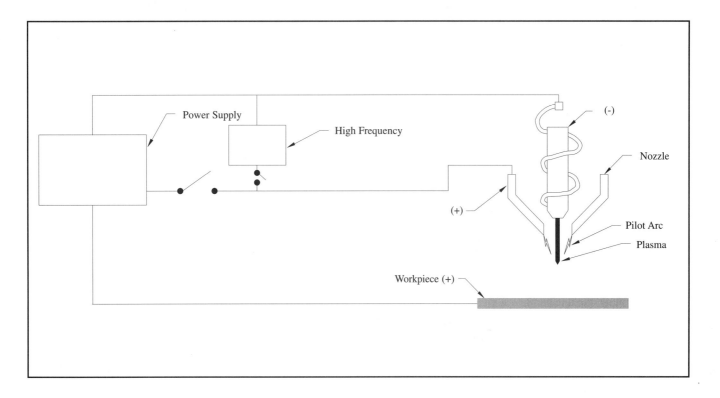

Figure 17:4
High frequency circuit ionizes the plasma and creates a pilot arc.

As the nozzle approaches the workpiece, the negatively-charged pilot arc is attracted to the positively-charged workpiece. An electronically-controlled sensor switches off the high frequency for the pilot arc. Gas ionization is then sustained from the main DC power supply.

As the heated plasma arc strikes the workpiece, the plasma arc melts the metal. Then the high pressure gas flow forces out the molten metal and pierces the workpiece. After the piercing occurs, the motion machine is activated and the metal is cut according to the programmed shape.

Plasma Processes

A. Conventional Plasma Cutting

Conventional plasma generally uses either air or nitrogen to cool and create plasma. This system is used primarily for hand held operations.

B. Dual Gas Plasma Cutting

Dual gas plasma uses one gas for creating plasma for cutting; the other gas is used to shield the cutting edge from the atmosphere. Various gas combinations are used to create the desired results. See Figure17:5.

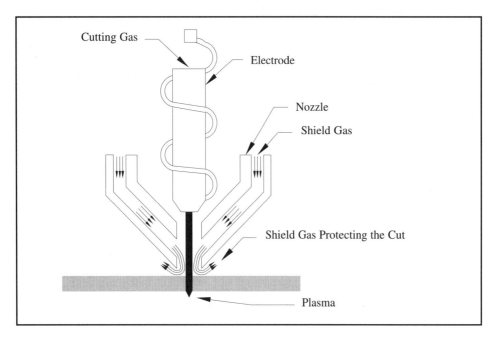

Figure 17:5
Dual Gas Plasma Cutting

C. Water Shield Plasma Cutting

Water shield plasma uses a similar process as dual gas plasma cutting. Instead of using a gas for shielding, it uses water to achieve better cooling for the nozzle and workpiece. This is especially useful for cutting stainless steel.

D. Water Injection Plasma Cutting

In water injection plasma cutting, a single gas is used. Water is injected into the plasma arc and increases the energy density of the gas as it leaves the nozzle. This process produces better quality cutting for many materials.

E. Precision Plasma Cutting

Precision plasma cutting machines use a special nozzle where a high-velocity mixing chamber equalizes plasma flow pressure. See Figure 17:6. A high-flow vortex or a magnetic field surrounding the plasma arc causes it to spin rapidly and narrows the plasma arc at the tip of the electrode. This fast-spinning plasma results in a finely defined beam that cuts a narrow kerf with a perpendicular edge.

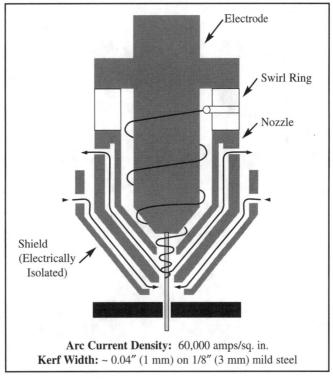

Electrode

Swirl Ring

Nozzle

Shield
(Electrically
Isolated)

Arc Current Density: 60,000 amps/sq. in.
Kerf Width: ~ 0.04″ (1 mm) on 1/8″ (3 mm) mild steel

Courtesy Hypertherm

Figure 17:6
Precision Plasma Cutting Nozzle

Manufacturers are increasingly using precision plasma cutting machines to produce parts. For many application it is an efficient cutting method. See Figures 17:7 and 17:8.

Courtesy Koike Aronson

Figure 17:7
Precision Plasma Cutting

Figure 17:8
Parts Cut With Precision Plasma

Difference Between Regular Plasma and Precision Plasma Cutting

Regular plasma cutting tends to leave a bottom residue of recast metal that is sometimes difficult to remove. Also in plasma cutting the electrode and the nozzle require frequent replacement. Besides these problems are two major concerns with conventional plasma cutting: beveled cuts and double arcing. See (Figure 17:9).

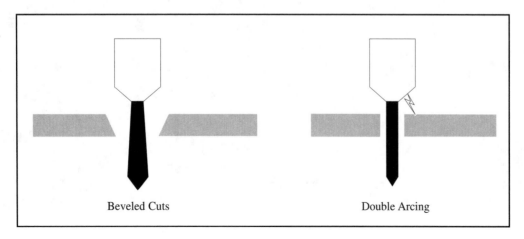

Beveled Cuts   Double Arcing

Figure 17:9
Conventional Plasma Problems

A. Beveled Cuts

In regular plasma cutting the jet stream is difficult to control and produces beveled cuts, particularly with materials under 1/4″ (6 mm). The bevel increases as the material gets thinner, and the kerfs can vary considerably during the life of a nozzle.

B. Double Arcing

Double arcing tends to occur in conventional plasma cutting. Double arcing hinders the cutting process and causes premature loss of the nozzle and the electrode.

Precision plasma machines overcome these deficiencies by using a high-velocity mixing chamber, aided by either a high-flow vortex or a magnetic field, to stabilize the pressure and to swirl the gas. The results produce a straighter kerf, minimum plasma deviation, and reduction of the chances for double arcing. Double arcing can also be minimized by using an isolated shield to protect the nozzle. In addition, with precision plasma cutting most bottom dross problems of the workpiece are greatly reduced. Some mild steel parts produce some slag; however, the slag can often be easily removed by hand.

Conventional plasma machines can cut material up to six inches thick (152 mm). Precision plasma machines cut primarily thin materials, under 5/8″ (15.86 mm).

Materials for Plasma Cutting

Plasma cutting requires electrically conductive material. Lasers, in contrast, because they use highly focused light, can cut both electrical conductive and non-conductive materials.

Heat Distortion and Heat Affected Zone

The workpiece absorbs much of the heat energy of conventional plasma cutting machines, thereby creating a greater heat affected zone. This can cause deformation. With precision plasma, most of the heat is used in the cutting process, resulting in considerably less workpiece heat. In a comparison study between distortion for precision plasma cutting and laser cutting, the results showed minimal differences.

Accuracy

The repeatability of the X-Y-Z movements and the vibrational stability of the machine are critical for cutting accuracy. The distance of the torch to the workpiece also affects part accuracy. If the torch is too high, it can create a long arc and put excessive heat into the material and cause it to have a beveled edge.

Precision plasma machines have much greater accuracy than conventional plasma machines. Laser cutting machines are still more accurate, but the gap is narrowing between lasers and precision plasma machines.

The general tolerances for production cutting with precision plasma is +/- .010″ (.254 mm). A single part can be cut within +/- .004 (.1 mm). The angularity of a cut on 3/8″ (9.52 mm) is +/- 1 to 3 degrees; below 3/16″ (4.76 mm) is +/- 0 to 1 1/2 degrees. Accuracy is also determined by the speed of the cutting machine.

Consumables

The electrodes and nozzles in plasma cutting need to be replaced periodically. Improvements have been made to prolong the life of these parts. However, the wear of these parts need to be considered in overall machine operation costs.

Torch-to-work height affects the life of the nozzle. Placing the plasma torch too close to the workpiece can cause premature damage to the nozzle tip in an unshielded torch.

Plasma and Shield Gases

Various plasma and shield gases are: air, nitrogen, argon and hydrogen, oxygen, and oxygen with nitrogen. Air is a commonly used plasma gas for systems under 200 amps. Nitrogen or oxygen is used for water injection cutting. A mixture of argon and hydrogen can be used for cutting thick stainless steel and aluminum sheets to provide better edge appearance. Oxygen is used for cutting carbon steels.

Several factors determining which gases to use: the metallurgical effects of the gases, cutting speed, cut quality, environment concerns, effects on consumables, and cost of gases.

Advantages and Disadvantages of Plasma Cutting Systems

The primary advantage of plasma cutting is the lower initial investment for the cutting system. Plasma cutting machines are less expensive than precision plasma cutting units, and precision plasma machines cost less than lasers.

Although lasers cost more, they provide a more accurate cut. Lasers also have the advantage of not being limited as plasma machines to cutting only electrical conductive materials.

Plasma and Turret Punch Presses

Machines are available that combine the efficiency of the turret punch press with plasma cutting. The advantage of a punch press is that with one stroke it produces the desired shape. The advantage of plasma cutting is that it can produce various shapes without expensive tooling costs. Machines capable of punching and using plasma for cutting are suitable primarily for high production runs. See Figures 17:10 - 17:12.

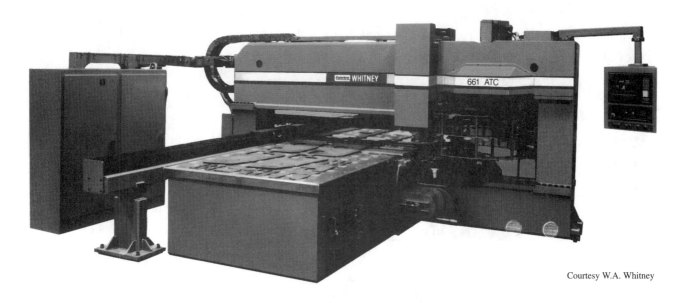

Courtesy W.A. Whitney

Figure 17:10
Combination Plasma and Turret Punch Press

Courtesy W.A. Whitney

Figure 17:11
Oxygen Plasma Cutting Head for Turret Punch Press

Courtesy W.A. Whitney

Figure 17:12
Automatic Tool Changer

Beside stamping out flat parts, turret presses offer another advantage. They can do both flat stamping and forming operations as shown in Figure17:13.

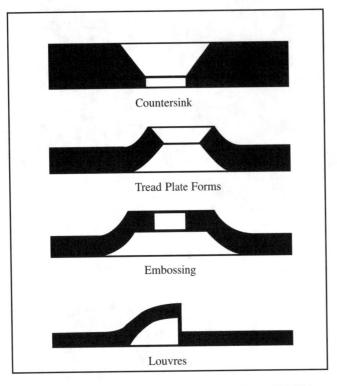

Courtesy W.A. Whitney

Figure 17:13
Possible Forming Operations With Turret Presses

Other Applications for Plasma

Plasma cutting utilizes a transferred-arc mode. The negative arc is transferred to the positive workpiece, causing the plasma to heat the material. The electrical conductivity of the workpiece is essential for this system to work.

Plasma machines can also operate in a non-transferred-arc mode. The nozzle contains an electron-collecting anode which creates energy. The heated gas can then be transferred to non-conductive surfaces. This system is used for surface treatment; cladding, as with spraying powdered oxides and carbides on cold workpieces; and for welding non-conductive surfaces, such as ceramics.

Plasma cutting devices can also be mounted on a robot and perform 3D cutting. Devices are available to maintain precise torch height while cutting.

Understanding Plasma Cutting

Plasma cutting, whether conventional or precision, is a fast, economical way to produce parts. Manufacturers should first understand the process, and then determine if this or another process produces the parts more effectively.

Unit 9

Waterjet and Abrasive Waterjet Machining

Notes

18 Waterjet and Abrasive Waterjet Machining

Manufacturers, in their quest for productivity and quality, are searching for alternate ways to cut various materials. Increasingly, they are turning to waterjet and abrasive waterjet machining.

Often the conventional machining method is to either saw cut or plasma cut the material, then machine the finish part. Abrasive waterjet machining often eliminates this secondary operation. Extremely powerful abrasive waterjets are capable of cutting through 3-inch tool steel at a rate of 1.5 inches a minute, 10-inch reinforced concrete at over 1 inches a minute, and 1-inch plate glass at 20 inches a minute.

Fundamentals of Waterjet and Abrasive Waterjet Machining

Modern use of waterjet machining technology dates to 1968. Dr. Norman C. Franz, Professor of Forestry of the University of British Columbia, was issued a patent for developing a high-pressure waterjet cutting system. See Figures 18:1 and 18:2.

Courtesy Flow Systems

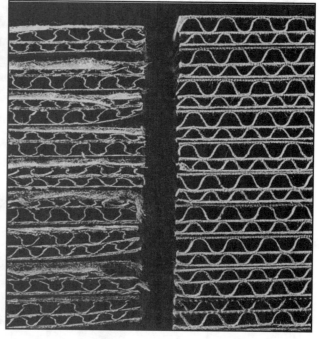

Courtesy Flow Systems

Figure 18:1
Cutting With High-Pressure Water

Figure 18:2
Comparison of Cutting Corrugated Boxboard: With Mechanical Knife (Left), With Waterjet (Right).

A modern waterjet system receives regular low-pressure water. To prevent premature failure of intensifier pump seals, system check valves, and the nozzle orifice, the water is filtered by one, two, or sometimes three series of filter banks. Water free from particles and minerals increases the life of the internal mechanisms of the waterjet system.

This filtered water proceeds through a high-pressure intensifier. The intensifier can increase the pressure from 20,000 to 60,000 psi, and accelerate the water to two to three times the speed of sound.

The pressurized water is transferred through high-pressure tubing to a cutting head. The high-pressure water passes through a small orifice, .003″ to .023″ (.08 mm to .6 mm), in the cutting head. After the water pierces the material, the used water is collected into a tank or a compact traveling catcher unit. The material is cut when the water pressure exceeds the material's compressive strength. The cutting head can be stationary, or it can travel from two to six axes. Most machines travel in the X and Y directions as shown in Figures 18:3 and 18:4.

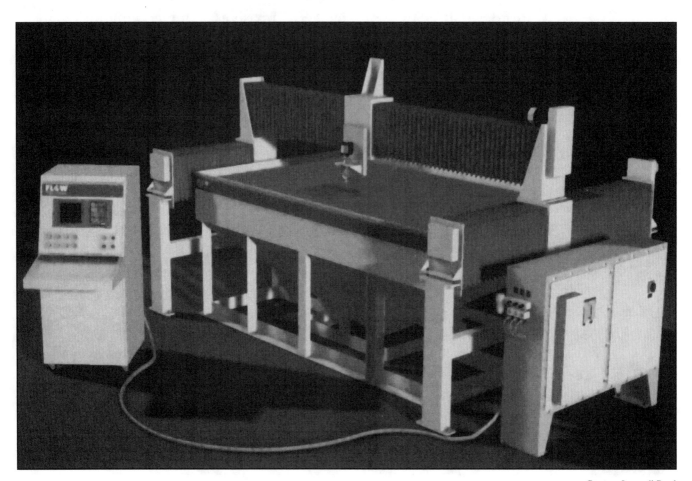

Courtesy Ingersoll-Rand

Figure 18:3
Gantry Type Waterjet Cutting Systems

Figure 18:4
Gantry Type Waterjet Cutting Systems

Introducing Abrasive Into the Waterjet

In 1983, Flow patented a process introducing abrasives into a high-pressure water system. The company then developed the first commercial abrasive waterjet system. See Figure 18:5

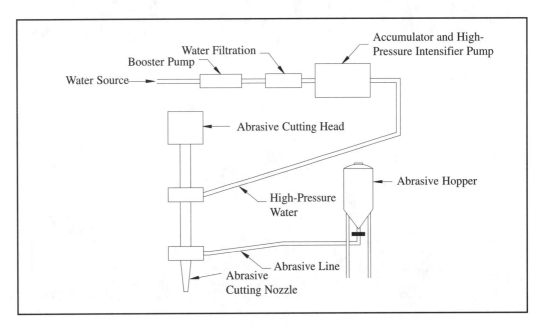

Figure 18:5
The Abrasive Waterjet Cutting System

With abrasives added to the high-pressure water stream, a wide variety of metals and non-metals can be cut. See Figures 18:6 and 18:7. Metal has been cut up to 12″ (305 mm) thick with abrasive waterjet.

Courtesy Flow Systems

Figure 18:6
Abrasive Waterjet Cutting

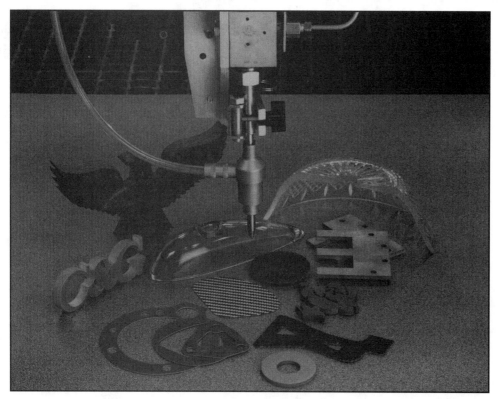

Courtesy Ingersol-Rand

Figure 18:7
Parts Cut With Abrasive Waterjet Cutting

Abrasives for Cutting

Some of the abrasives used for abrasive waterjet cutting are: garnet, silica, aluminum oxide, crushed glass, copper slag, silicon carbide, and silicon nitrite in mesh sizes from 36 to 150. Garnet in the 80 to 100 mesh sizes is primarily used for cutting. Generally, the harder the abrasive, the faster the cutting. However, harder abrasives increase nozzle wear. Figure 18:8 shows abrasive waterjet cutting through various materials.

Courtesy Jet Edge

Figure 18:8
Abrasive Waterjet Cutting Through Various Materials

Kerf width with abrasive waterjet machining is usually from .030″ to .100″ (.76 mm to 2.5 mm). The kerf is usually .005″ (1.3 mm) wider than the mixing tube diameter. Surface finishes range from 80 μin Ra to 250 μin Ra. The finer the abrasives, the better the surface finish; however, the fine abrasives decrease cutting speed.

The Abrasive System

The abrasive system includes the abrasive hopper, type and size of abrasive, abrasive metering valve, and delivery lines. The hopper is designed so abrasives flow smoothly without clogging. The metering valve regulates the amount of abrasives used, turns the flow of abrasives on and off, and purges the feed line of any water when the abrasive waterjet is not in use.

The Abrasive Cutting Head

The abrasive cutting head consists of:

1. A sapphire, ruby, or diamond orifice to emit high-pressured water.

2. A valve actuator to control the high-pressured water.

3. A chamber where regulated amounts of abrasives from the metering valve are mixed.

4. The nozzle (also called the mixing tube) through which abrasives and high pressure water flow.

To cut accurately, nozzles must be monitored and periodically replaced because of wear from the abrasives flowing through them. Much improvement has occurred in nozzle life from earlier days. Some nozzles are made from ceramic to prolong their life. Various nozzle diameters are used to produce the desired stream configurations for optimum cutting.

The main parameters for cutting are: pressure, abrasive flow, nozzle diameter, nozzle standoff distance, material type and thickness, orifice diameter, abrasive type, and cutting speed rate.

Motion Control Systems

Because the cutting head is relatively light, the motion control systems for waterjet cutting need not be massive. Various motion control systems are employed, from systems with multiple cutting heads to six-axis systems and robotic devices. One manufacturer mounted an abrasive waterjet system over a large variable speed rotary table. That system cut out the center slug of forged round blanks. See Figures 18:9 and 18:10 for various cutting systems.

Courtesy Jet Edge

Figure 18:9
Multiple Head Waterjet Cutting System

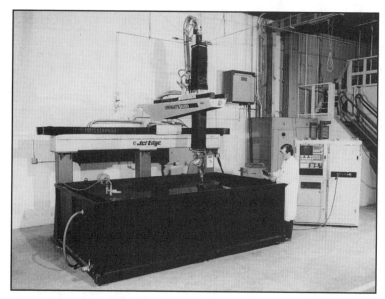

Courtesy Jet Edge

Courtesy ABB I-R Waterjet Systems L.L.C.

Figure 18:10
Multi-Axis Waterjet Cutting Systems

One manufacturer has enclosed its entire system. It enclosed a six-axis robot that is suspended vertically, and whose longitudinal travel within the cutting chamber provides a seventh axis. See Figure 18:11.

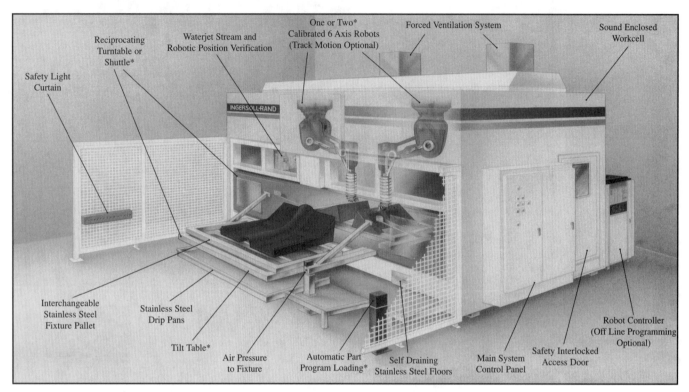

Courtesy Ingersol-Rand

Figure 18:11
Six-Axis Robotic Waterjet Cutting System

Catcher System

Waterjet cutting requires a system to catch the high-pressure water and abrasives that exit from the material being cut. A two-axis system commonly uses a large water-filled tank for a catcher system. The spent abrasives are collected in the bottom of the tank and then removed.

A catcher system is also necessary because the escaping abrasive waterjet can have energies up to 60 HP at the nozzle with only 75% of this energy being used while cutting. Some tanks are filled with steel balls to deflect the spent energy.

Also used are outside disposal system, where the spent water and abrasives are pumped into a drum. The heavier abrasives sink to the bottom of the disposal drum, while the water returns to the tank to help divert the forces of the escaping abrasives and water. Generally, the overflowing water is discarded.

When cutting with a multi-axis system and the opposite side of the object being cut is exposed to the escaping abrasives, there must be a catcher system to prevent the escaping abrasives from damaging the opposite side. Usually a can containing stainless steel balls is used to dissipate the spent energy. As the steel balls wear, more are added.

The next chapter examines the various uses of waterjet and abrasive waterjet cutting.

19 Profiting With Waterjet and Abrasive Waterjet Cutting

Many companies have discovered profitable uses for waterjet and abrasive waterjet cutting machines. Thousands of waterjets are now operating in production environments.

Materials Cut With Waterjet

Waterjets, apart from using any abrasives, can cut many materials. Even thin metals have been cut with high water pressure. See Figure 19:1.

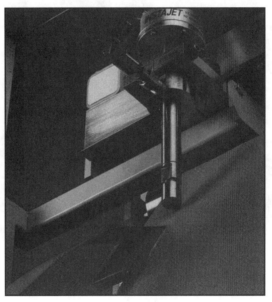

Courtesy Flow Systems

Figure 19:1
Slicing a fifty-pound liner board, dust-free, at 6,000 feet per minute.

Listed are some materials that can be cut with waterjet:

Thin Aluminum	Lead	Fiberglass
Plexiglass	Kevlar	Circuit Boards
Corrugated Boxboard	Wood	Rubber
Graphite Composite	Carpet	Food Products

Materials Cut With Abrasive Waterjet

Since abrasive waterjet cutting is a basic mechanical cutting operation, it can cut any type of material that is softer than the abrasives used. See Figures 19:2 and 19:3.

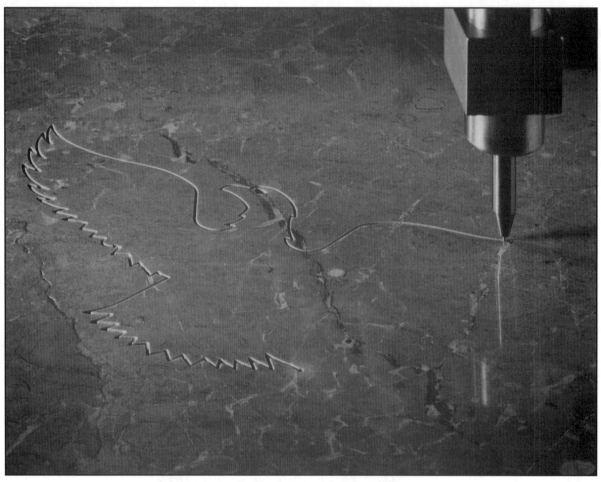

Courtesy Ingersol-Rand

Figure 19:2
Cutting Marble With Abrasive Waterjet

Listed are some materials that can be cut with abrasive waterjets:

Aluminum	Brass	Copper
Lead	Magnesium	Carbon Steel
Alloy Steel	Stainless Steel	Tool Steel
Cast Iron	Inconel	Titanium
Plastics	Bonded Materials	Fiberglass
Ceramics	Lexan	Rubber
Composites	Graphite/Epoxy	Glass

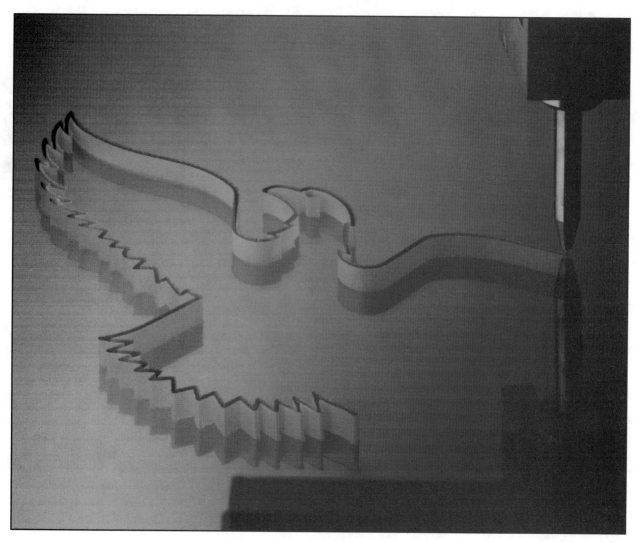

Figure 19:3
Cutting Glass With Abrasive Waterjet

Accuracy

High-accuracy positioning devices can cut within +/- .005″ (.127 mm) and closer. Production cutting uses a tolerance of +/- .005 to. 015″ (.127 to.38 mm). Stock thickness greatly affects cut accuracy.

The top surface of thick cuts tend to be smoother than the bottom, and the bottom cuts tend to be tapered and have jet-induced striations. Slowing down the cutting speed improves the surface finish.

Safety

Waterjet can cut flesh and bone easily; therefore, operators must be protected from the high-pressure forces. Likewise the noise level on some machines, from 80 to 120 DB, poses a hazard and require ear protection. To lower noise level, some machines cut under water which reduces the noise level to under 75 DB, some surround the cutting head with a noise suppression device, and some enclose the waterjet cutting system as shown in Figure 19:4.

Courtesy Ingersol-Rand

Figure 19:4
Enclosed Waterjet Cutting System

Since waterjet or abrasive waterjet cutting creates no fumes, no danger results from toxic fumes. However, the escaping abrasives, particularly when piercing a hole, escape to the atmosphere and coat surrounding equipment. For this reason some manufacturers enclose the cutting system.

Disadvantages of Waterjet and Abrasive Waterjet Machining

A. Frosting From Abrasive Waterjet Cutting

During some applications frosting can occur from abrasive waterjet cutting. When frosting is unacceptable, a dehazer device can be attached to the cutting head. With this device low-pressure water surrounds the cutting stream and prevents the high-pressure abrasive stream from frosting the material. The dehazer is particularly useful when cutting glass, brass, marble, or other shiny materials. Its added benefit is it lowers the noise level and the airborne abrasive dust. It also increases the standoff with no appreciable change in kerf.

B. Slower Speed Rates and Higher Costs Than Plasma and Lasers

Plasma cutting machines and lasers cut faster and cheaper, but both leave a heat affected zone. However, thick materials limit laser cutting ability.

C. Catchers Needed With Multi-Axis Cutting

Because of the high-velocity abrasives leaving the cut, a catcher system needs to be behind the cutting head to prevent part damage. Lasers generally do not have this problem.

D. Large Cuts Become Stratified

Thick cuts tend to produce stratified surfaces. The waviness of the cut is greatest on the bottom of the cut. In contrast, the electrode of wire EDM travels along the entire surface of the cut and produces an exceptionally smooth surface, even for the first cut.

Advantages of Waterjet and Abrasive Waterjet Machining

A. Material Savings

Much material waste often occurs with sawing and milling. Since abrasive waterjet cutting produces a small kerf, parts can be nested for maximum material utilization.

B. No Special Tooling Required

Waterjet permits parts to be cut without special tooling or cutters. The desired shape is programmed and downloaded into the waterjet system. When a design needs to be changed, the program can be altered with little production downtime. Also, when material or workpiece thickness changes, only the cutting speed parameters need to be changed. This provides great flexibility and permits just-in-time machining.

C. Moisture Absorption Not a Problem

Materials such as corrugated board, paper, and carpet absorb little moisture because of the high velocity of the waterjet system. Any dampness that may occur rarely affects the product.

D. Focusing the Waterjet is Not Critical

Unlike lasers, where focus is critical, the standoff distances of waterjet, generally under one inch (25 mm), produce little cut variation. However, optimum distance is usually twice the diameter of the mixing tube's ID.

E. Simple Fixtures Required

Since the waterjet cutting forces are extremely low, from 3 to 5 lbs (1.36 to 2.27 kg), simple fixtures can hold parts. However, materials with internal stresses must be secured to prevent the parts from moving while being cut.

F. Entry Hole Not Required

Abrasive waterjet produces its own entry holes. Parts with multiple cavities can be machined without any other premachining.

G. No Heat Affected Zone or Microcracking

Waterjet and abrasive waterjet machining create no heat-affected zone at the cut edge, as do EDM, lasers, or plasma cutting machines. See Figure 19:5.

Courtesy Flow Systems

Figure 19:5
Cutting Titanium With No Heat Affected Zone

The cold-cutting operation protects the microstructure of the cut surfaces, and prevents edge work hardening, microcracking, and heat distortion. One company discovered that their titanium dental implants cut with lasers suffered from microscopic heat-stress cracks. These cracks permitted harmful bacteria to develop in the implants. Abrasive waterjet solved the problem.

H. No Fumes

Lasers require a system to remove cutting fumes. Neither waterjet or abrasive waterjet cutting produces fumes.

I. Eliminates Some Difficult Cutting Problems

To drill through carpet is an extremely difficult operation because the carpet tends to wrap around the drill. Waterjet machining eliminates this problem. Laminated products can also be cut without separating the laminations.

Traditional cutting of composites often damaged them with heat, fraying, or edge delaminating. In addition, diamond or carbide-tipped routers, bandsaws, or abrasive wheels cut slowly. Abrasive waterjet solves these composite cutting problems.

J. Burr Free Cutting

The abrasives leave no burrs or slag on the material.

K. Easily Attached to Robots

Since omnidirectional waterjets and abrasives waterjets cut with forces under 15 lbs (6.8 kg), they can be readily adapted to robots for automated operations.

Cutting With Multiple Heads and Stacking Materials

The small thrust of waterjets allows multiple heads to be mounted on a motion system to multiply production. Also thin materials can be stacked. See Figure 19:6.

Courtesy Flow Systems

Figure 19:6
Dual Heads Cutting Stacked Materials

Glass Sculpturing

Abrasive waterjet cuts plate glass without causing it to crack. To etch designs in the glass, a sandblaster is used. The glass is covered with tape (one method is to use 3M, 6 inch (152 mm) tape, #280), and then a pattern is drawn on the tape. The part that is to be deeply etched is first cut out with a knife. A protected worker in a booth uses a portable sandblast unit with 100 grit garnet to blast the exposed glass. The next surface to be etched is then cut. This procedure is repeated until the desired etched surfaces are achieved.

Turning With Abrasive Waterjet

Workpieces can be turned with abrasive waterjets. As the workpiece rotates, the abrasive waterjets removes the material. With this process glass can be turned. Even treads can be put on glass.

Waterjet and Abrasive Waterjet Capabilities

Dramatic results have been achieved with waterjet and abrasive waterjet cutting. They have efficiently sliced food, as well as cut composites, hardened steel, and marble. See Figures 19:7 and 19:8.

Courtesy Flow Systems

Figure 19:7
Meat Being Slit for Packaging

Courtesy Flow Systems

Figure 19:8
A limited edition of an automobile bumper being modified
with waterjet and requiring no further hand finishing.

As has been shown, high-pressure waterjets and abrasive waterjets can be used in many ways. Manufacturers should explore their unique capabilities to increase productivity.

Unit 10

Lasers

Notes

20 Fundamentals of Lasers

Lasers: The Revolutionary Concept

Lasers can be found in all sorts of human endeavors. Lasers drill diamonds, teeth, and steel. Lasers weld computer chips, autos, and detached retinas. Trained eye surgeons strike the retina with a blue argon laser beam which bonds the detached retina. Before this break-through technology, traumatic freezing and surgical procedures were used.

Lasers align machines, read merchandise codes, print documents, play audio discs, and send images and sound over thin fiber optic cables thousands of miles. Coherent laser light beams can be as high as one million times more powerful than earth's sun ray beams. The following chapters on lasers will include cutting, welding, cladding, drilling, hardening, marking, and other procedures related to laser manufacturing.

Laser Cutting

Increasingly, laser cutting is becoming the first choice for many manufacturers. Laser cutting machines range in power to 5000 watts and cut with speeds from 1 to 600 inches (25 mm to 15,000 mm) per minute. See Figures 20:1 and 20:2.

Courtesy Trumph Laser

Figure 20:1
A CO$_2$ Cutting Laser

Figure 20:2
Production Laser Cutting

The accuracy, speed, and minimum workpiece distortion of lasers have altered the manufacturing of many parts. On a worldwide basis for industrial lasers, over 40% of the lasers are used for cutting, about 20% for welding, and the remainder for drilling, hardening, cladding, marking, and others.

In 1960, Ted Maiman of Hughes Corporation developed the first laser from a ruby crystal. The intensity of the light surpassed anything known before that time. The first production laser was installed in 1965 when a ruby laser drilled small holes in diamonds. Today, thousands of lasers perform various operations in manufacturing and job shops around the world.

Lasers are extremely cost effective because they eliminate expensive hard tooling. Lasers can economically produce round holes, square cutouts, radii, tapers, and undercuts to any imaginable shape, without expensive tooling costs as with turret or power presses. Looking at the parts in Figure 20:3, one could imagine the expensive tooling that would be needed if lasers or other cutting systems were unavailable.

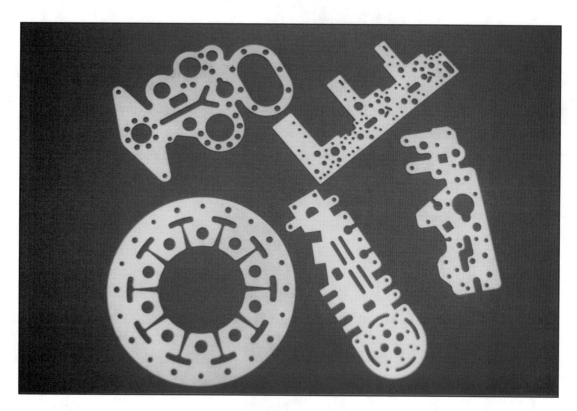

Figure 20:3
Various Shaped Parts Cut With Lasers

To produce the various shapes, a desired shape is programmed on a CAD system, postprocessed to convert CAD design language to CNC language, and then downloaded into a laser. In most cases, the laser cut part requires no further finishing. Understanding lasers can result in substantial increases in productivity. See Figure 20:4.

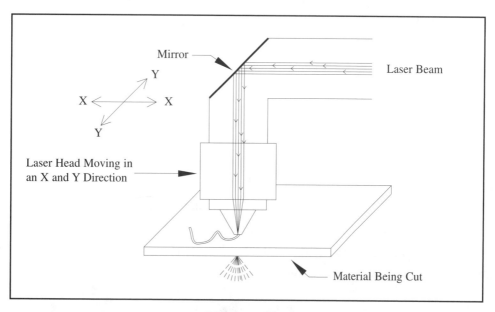

Figure 20:4
Laser Cutting Machine

How Lasers Work

Laser, an acronym for Light Amplification by Stimulated Emission of Radiation, receives its energy from the amplification of light. A laser consists of a lasing medium such as gas, crystal or liquid; an energy source to stimulate an emission of photons from the medium such as electricity, flashlamp, or another laser; and mirrors to provide optical feedback and amplification. Some of the light energy is allowed to escape at one end of the laser, and then the beam can be directed by mirrors or fiber optics to the workpiece.

Resonator

A resonator is the unit which creates the high energy light beam which passes through optics. In an axial flow CO_2 laser resonator, voltage is applied to an anode and a cathode in a tube filled with helium, nitrogen, and CO_2 gases. In a 1500 watt laser, up to 20,000 volts of DC current excites the gases—raising the energy level of the lasing medium. When the lasing medium returns to its unexcited state, a photon of energy is given off. This photon collides with another excited atom of the lasing medium and produces another photon. In millionths of a second, a laser beam is created as the photons bounce back and forth between the resonator mirrors.

This "stimulated emission of radiation" creates a chain reaction resulting in laser light. Helium inside the resonator acts as a cooling gas. The laser gases are circulated by means of a pump and cooled by a chiller.

The invisible laser light is monochromatic; it consists of only one color and one wavelength. The wave length travels in only one direction and coherently; i.e: the wave lengths are in phase with each other. Laser light photons travel an extremely narrow path. For example, a laser beam was aimed at a mirror on the moon. After making a round trip of half a million miles, the reflected laser light was picked up on earth.

Laser Mirrors

With resonators in the shape of an "U," the laser light inside the resonator bounces back and forth against reflective mirrors until it increases to an intensity that can pass through a partially reflective optic (PR mirror). These PR mirrors are typically from 35 to 80% reflective with the remaining transmissive. Light which has not passed through the partially reflective mirror bounces back, hitting the nitrogen gas which collides with the CO_2 gas. This collision then creates more photons to hit the mirrors. See Figure 20:5.

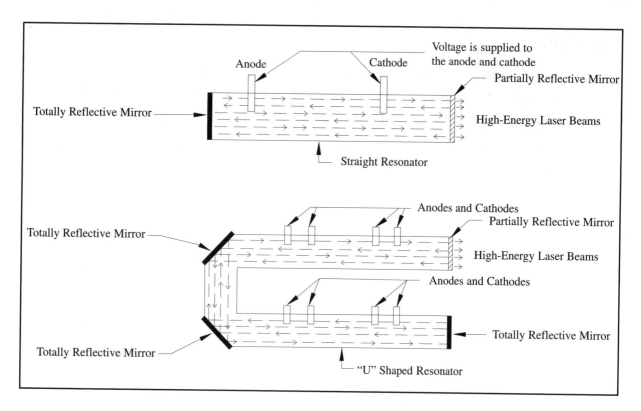

Figure 20:5
Straight Resonator and Resonator Shaped as a "U" to Produce Laser Beams

The high-energy beams of light eventually pass through the partially reflective mirror and are directed by totally reflective mirrors (TR mirrors) to the laser nozzle as shown in Figure 20:6.

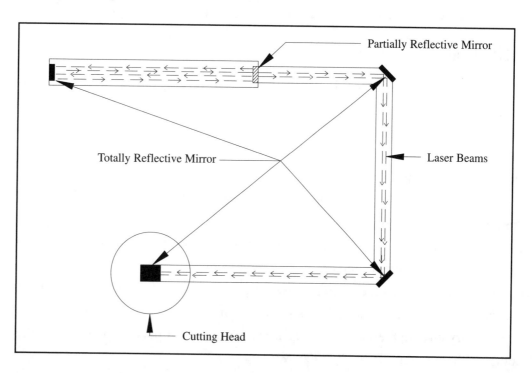

Figure 20:6
The Transfer of Laser Light to the Cutting Head for a Flying Optic Laser

Laser Optics

A lens at the cutting nozzle focuses the laser beam as illustrated in Figure 20:7. The lens focuses the laser energy onto a small spot, which greatly increases its intensity. The thickness of the material being cut determines the size of the lenses.

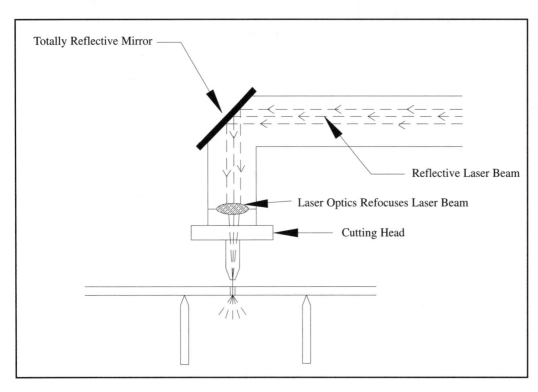

Figure 20:7
Laser Optics Focus the Laser Beam Which Creates Intense Energy of Light

Assist Gases

The laser beam is supported by an assist gas of air, argon, helium, nitrogen, or oxygen. When oxygen is used in metal cutting, it allows the laser to cut faster by assisting the laser beam to vaporize the molten material and to force out the molten material. Sometimes a stream of water or a water shield is used to prevent thermal effects when cutting with oxygen.

When inert gases such as argon or nitrogen are used, they aid only to drive out the molten material from the gap; they do not aid the cutting process. Nitrogen is used to cut stainless steel to provide an oxide-free edge, sometimes called "clean cut." The advantage of "clean cut," though slower than oxygen cutting, is it eliminates a cleaning process to remove oxide scale from the cut surface. This is particularly important when stainless steel parts need to be welded or are to be used in a sanitary environment. Inert gases are also used to cut plastics, wood and other nonmetallic materials. See Figure 20:8.

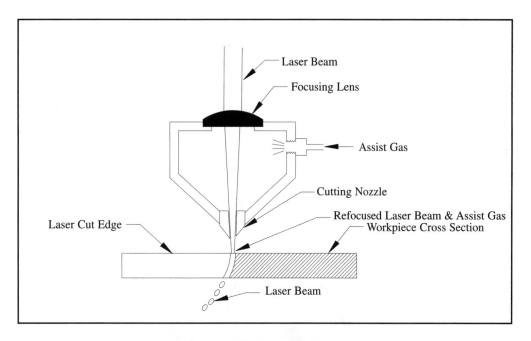

Figure 20:8
The Laser Beam and Assist Gas

The Laser Cut

High pressure assist gas quickly forces the molten material from the kerf. The molten material absorbs much of the heat; however, some heat is absorbed by the material, leaving a small amount of recast. This minimal resolidification of molten material is an important benefit of laser cutting, resulting in a minimal heat-affected zone and heat distortion.

The laser beam melts the material and helps to vaporize it as the assist gas forces out the molten mass. The combination of assist gas pressure and evaporation removes the molten material as shown in Figure 20:9. An exhaust fan and filter system removes the gases and fumes.

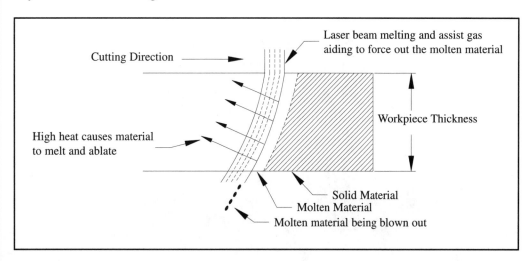

Figure 20:9
Laser Beam Melting the Material as the Assist Gas Forces Out the Molten Mass

Sensing Unit

On some lasers, a height-sensing unit maintains the focal point at a fixed distance below or above the workpiece, even though the surface of the workpiece may fluctuate. This assures cutting uniformity. To prevent nozzle damage, some lasers are equipped with height scanners and a collision protection device. See Figure 20:10.

Courtesy Lumonics

Figure 20:10
Laser Nozzle Equipped With a Beam Director for Crash Protection

Laser Safety

Lasers operate at high voltages. Only qualified personnel should work on these machines. Laser personnel must be informed and trained to avoid eye and skin damage from direct and reflected laser beams. Operators should wear safety glasses designed specifically for laser wavelengths. The sign on the cutting laser in Figure 20:11 states, "AVOID EXPOSURE: Visible and invisible laser radiation is emitted from this aperture."

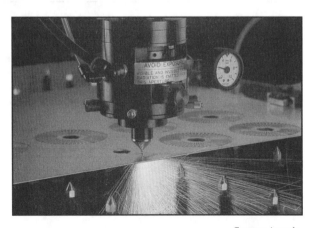

Courtesy Amanda

Figure 20:11
A Cutting Laser with a Safety Warning

Fumes From Laser Cutting

Surface applications, such as oil, create fumes when cut with lasers. An exhaust filter system will remove these fumes and suspended particles created from laser cutting. In addition, purchasers of laser processing systems should be aware that certain materials create toxic fumes that need to be removed.

21 Understanding Laser Cutting

Kerf Width

The round laser beam for cutting creates a narrow kerf width from .006″ to .010″ (.15 mm to .25 mm). As a result of the round cutting beam, all internal radii will have at least a .003″ to .005″ (.08 mm to .13 mm) radius, as shown in Figure 21:1. Generally, this creates no problem; in fact, corner radii strengthen parts.

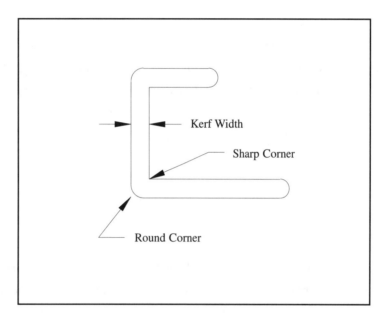

Figure 21:1
Kerf Width and Internal Radii

Material Distortion

Although some thermal distortion may occur from laser cutting, distortion comes primarily from relief of internal material stresses. As the material is cut, internal stresses are relieved. This relief causes parts to move. Cold rolled materials are particularly vulnerable to such distortion.

Heat-affected Zone

Laser cutting produces a small heat-affected zone. However, studies show that laser radiation and the reactive gas flow are absorbed mainly by the molten material and less by the walls of the cut. This absorption effect minimizes the heat-affected zone of the laser cut.

Edge Quality

Edge quality is determined by such factors as material thickness, material type, part configuration, type of assist gas and its pressure, gas nozzle design, focus position, type of lens, optic cleanliness, beam quality, and cutting speed. The thinner the material and the finer the steel grain, the better the edge quality. High alloy steels such as spring steel and tool steel have very good edge quality.

Thicker materials exhibit substantial reduction in edge quality; however, for many applications, edge quality is more than satisfactory. In cases of unsatisfactory edge quality, conventional machining, water jet, or wire EDM should be used.

Test Cuts

To determine maximum laser efficiency and to produce a satisfactory edge surface, various factors need to be considered, such as material type and its surface condition, feed rates for straight and corner cutting, cutting gas pressure, and the amount of laser power. To achieve maximum laser efficiency, test cuts are usually made to set proper laser parameters. When material is supplied to laser shops, extra material for test cuts should be provided.

Reducing Costs

Speed of Lasers

The cutting speed of lasers is determined by the type of material, the material thickness, the assist gas, beam quality, and the required edge finish. Generally, the material type and thickness cannot be changed. For non-critical parts, allowing for a coarser edge surface allows the laser to cut faster.

Generally, oxygen cuts stainless steel faster; however, it leaves a dark edge with an oxide scale. This dark edge hinders welding. Though more costly, nitrogen can be used to cut stainless steel to produce a clean edge for welding.

Tolerances

The closer the tolerances, the slower the cutting speeds. On thin work, lasers can hold up to +/-.003″ (.08 mm); some claim +/-.001″ (.025 mm) and closer. However, whenever possible, more tolerances should be allowed, generally +/-.005″ (.13 mm).

Many materials have internal stresses. When these materials are cut with laser, the parts often move during the cutting process. This factor should be considered when stipulating tolerances.

Factors such as the acceleration and deceleration of the laser machine, part thicknesses, and the lagging effects of the laser cut, a slowing down and dwell time for sharp corners, is put into the program. Various slowing down and dwell times are used depending on material type and thicknesses. By allowing greater tolerances, the laser is permitted to cut faster.

Surface Condition

The surface condition of the material is important. Rust or scale on the surface impedes the cutting process and can cause irregular surfaces and excavations. Buying material such as pickled and oiled mild steel or cleaning the surface by sandblasting helps in solving the problem.

Beam Quality

Beam quality is one of most important factors for efficient laser operations. In laser cutting, the radiation power, radial distribution, and the circular pattern of the beam are important factors in maximizing the accuracy and effectiveness of the laser.

There are two main types of laser beams for cutting—the Gaussian TEM00 and the Doughnut TEM01 beams. Figure 21:2 shows the two different beams.

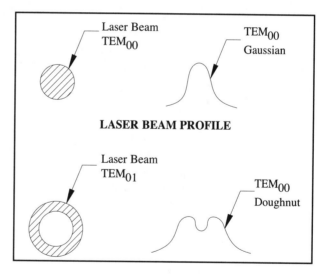

Figure 21:2
Two Types of Laser Beam Profiles for CO_2 Lasers

One method used for checking the laser beam is to allow the unfocused beam to melt a clear bar of acrylic plastic to a certain depth. To determine the laser beam cutting effectiveness, the burned shape pattern is analyzed.

Circular polarization of the laser beam has a strong effect on the cutting accuracy. To check for beam quality, eight slits should be cut in a piece of mild steel, each 45 degrees apart from the center. The bottom and top of the slits should be checked for any deviation. Cleaning, adjusting, or replacing a defective unit may be required to eliminate the deviations.

Beam Focal Length

To achieve maximum efficiency from the laser, the laser beam needs to be properly focused on the workpiece. Various focal lengths, 3.75, 5, 7.5, and 10 inches, (95, 127, 190, and 254 mm) are used for different thicknesses of material. The thicker the material, the larger the focal length. Improper focal length affects the cut quality and edge parallelism.

The waist of the focused hourglass shape of the laser beam is the area of greatest power concentration. It is important that the laser beam cuts in this area for maximum efficiency. If a small focal length lens is used to cut thick material, it will slow the machining process and affect edge parallelism. See Figure 21:3.

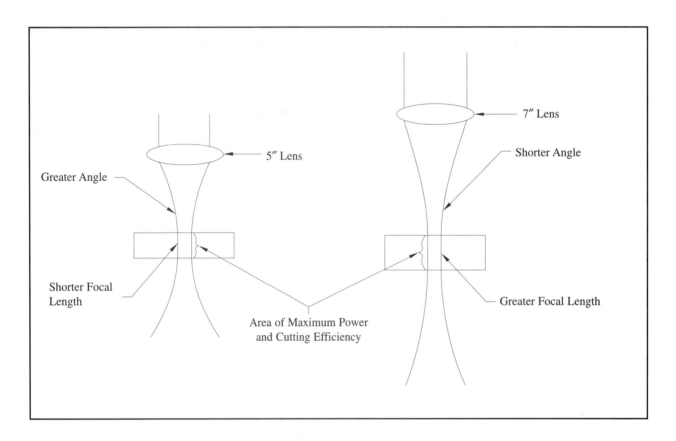

Figure 21:3
Effects of Focal Lengths

Quantity

Laser cutting can be cost effective even for one part. However, for making large orders, the more cost effective action is to seek the maximum number of parts from a sheet, rather than stipulating a certain number of parts. For example, rather than order 1000 parts, order maximum number from five sheets of material.

Redesigning Parts

With no tooling restriction, any shape can be programmed on the CAD system. The only limitation is the small internal radii, from .003 to .005" (.076 to .127 mm). Holes, notches, radii, and angles can be cut easily.

It is difficult when using hard tooling to produce thin relieving cuts. For example: after production, a part is to be bent 90 degrees. With laser cutting, a relief cut can be easily added to strengthen the bent part (See Figure 21:4).

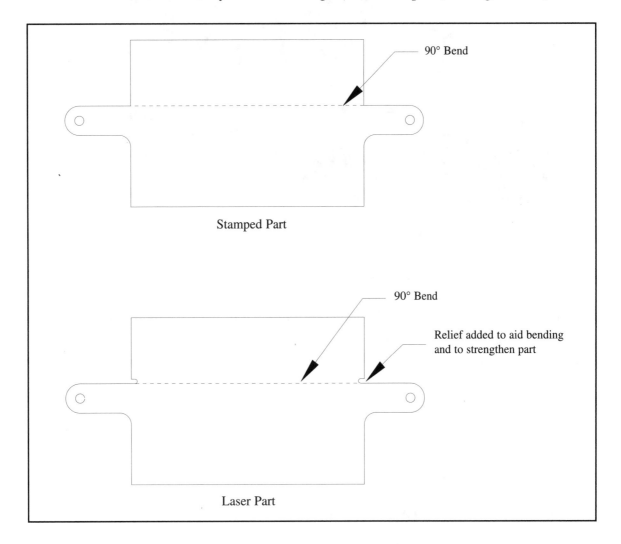

Figure 21:4
Capabilities of Adding Features Without Substantial Costs

Lasers and Turret Punch Presses

Turret punch presses have the distinct advantage of producing a desired shape with one stroke. A laser, however, must travel the entire contour. But lasers have the advantage of producing various small and large shapes without expensive tooling. Also, lasers can produce all desired parts immediately; whereas, turret punch presses may require lead time to produce the desired tooling. To increase productivity, some manufacturers have combined turret punch presses and lasers as illustrated in Figure 21:5.

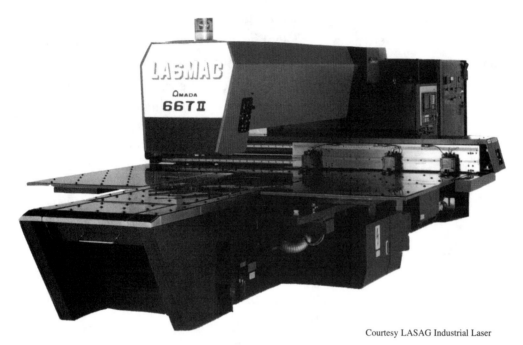

Figure 21:5
Laser and Turret Punch Press

Obviously, this combination of laser and turret punch press is for high-volume production. For small scale production, lasers have the distinct advantage of not requiring various kinds of tooling as do turret punch presses. Generally, when lot sizes increase, turret punch presses become more economical; when lot sizes are substantial, then conventional punch presses are more cost effective. Each cutting system has its advantages. A benefit of turret punch presses over lasers is their ability to do some forming operations like stamping reinforcing beads on sheet metal panels or countersinking holes.

Cutting Laser Increasingly in Demand

Increasingly, manufacturers are using lasers for cutting due to their speed, finish, accuracy and elimination of costly tooling. The next chapter examines other lasers and their applications.

22 Various Lasers and Their Configurations

Two lasers dominate the material-processing field—the CO_2 laser and the Nd:YAG laser.

How YAG Lasers Work

The Nd:YAG laser, (neodymium-doped yttrium aluminum garnet) is a solid-state laser. Solid state means that the laser beam is produced by a crystal. The laser rod, which has been doped with an optical pure material, can lase. The yttrium aluminum garnet, a synthetically grown crystal, is doped with neodymium atoms. In contrast, the CO_2 laser is a gas laser. The CO_2 gas laser uses a mixture of helium and carbon dioxide within a chamber to obtain laser energy instead of a solid rod.

Flashlamps within the Nd:YAG lasers excite the media, in a process called pumping, which produces the laser beam. In Nd:YAG lasers, a very small amount of the electrical power sent to the flashlamps is converted to laser energy; most of the power turns to heat. The heat is withdrawn from the laser by a coolant system.

As laser beams emerge from the laser rod, the beams strike the rear reflective mirror, then they reflect to the front of the partially reflective mirror. When beams of sufficient energy develop, they escape from the partially reflective mirror to form the laser beam used for production. See Figure 22:1.

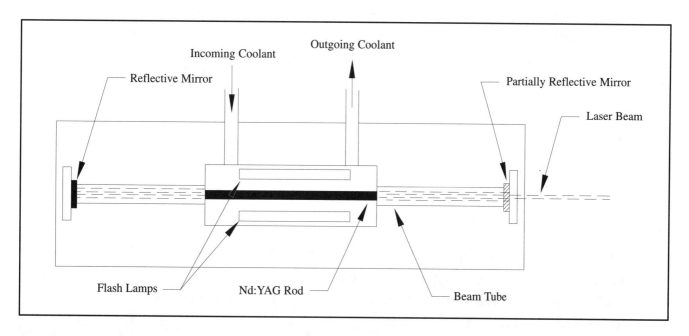

Figure 22:1
Nd:YAG Laser

Nd:YAG lasers come in continuous wave and pulsed mode. Flashing the lamp inside the laser produces the pulsing. Pulsed mode can increase laser peak power up to 500 times, and is particularly useful for drilling and welding.

Increasing Power for Nd:YAG Lasers

To increase the power for Nd:YAG lasers, individual Nd:YAG lasers are connected in series. Four 500 watt lasers connected in series produce approximately 2000 watts of power.

For greater cutting power, the CO_2 lasers increase their power to 3000 watts. The advantage of this higher power is faster processing rates for thicker materials. Also, the wave length of the CO_2 laser aids in cutting non-metals.

Various Lasers

The wavelength of the laser beam determines the various kinds of lasers. Listed are some of the various lasers and their wavelengths:

Lasers	Wave Length
CO_2 (carbon dioxide)	10.6
YAG (Yttrium aluminum-garnet)	1.6
CO (carbon monoxide)	6.0
HF (hydrogen floride)	2.7
CVL (copper-vapor lasers)	.5
Excimer	.193-.351

There are other solid state lasers such as neodymium doped glass lasers and yttrium lithium fluoride lasers. Previously, alexandrite and ruby lasers showed promise in drilling; however, Nd:YAG lasers have replaced them. Though they are not frequently used, Nd:glass lasers are used for their low power and low frequency. Since Nd:YAG is the predominant laser next to the CO_2 laser, its benefits will be examined.

Benefits of Nd:YAG Lasers

1. Fiber Optics

One of the great benefits of Nd:YAG lasers is the ability of the laser beam to be used with fiber optics as shown in Figure 22:2. Some Nd:YAG lasers are capable of using a fiber optic cable that is over 600 feet (183 M) long. This allows the user to reach into remote and difficult locations.

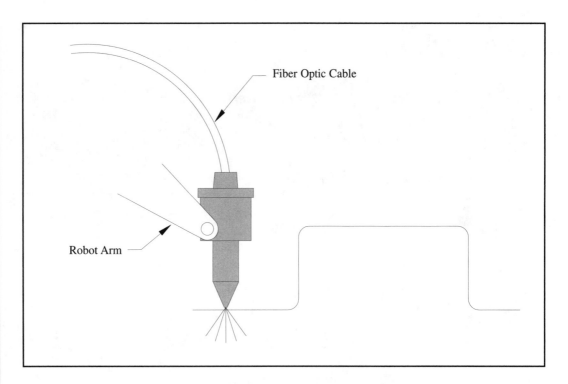

Figure 22:2
Nd:YAG Laser With Fiber Optics Trimming Sheet Metal

This ability of lasers to use high-power-density beams through fiber optics allows for on-line production with robots. This fiber delivery system also allows multistation use by sharing the beam. This "time-sharing" of the laser beam is commonly used to minimize distortion during the welding process. Time-sharing also allows the laser to be used while loading and unloading.

With an articulated-arm or a five-axis robotic system, Nd:YAG can use a conventional positioning system with its fiber optics. However, CO_2 lasers generally deliver their beams to the workstation by means of an expensive specially engineered mirror system.

2. Smaller Focusing Beam

Since a Nd:YAG laser beam is about one tenth the wave length of CO_2 lasers, it can focus to a very small diameter. This produces a higher energy concentration, making it ideal for drilling.

Excimer Lasers

Excimer lasers were first developed in 1975. The niche for excimer lasers is micromachining of ceramics, glass, polymers, plastics, metals, and diamonds. These lasers have been used for biomedical flow orifices, hybrid microelectronics, optoelectronics, diamond substrate machining, micromotors, fine wire stripping, and small optical apertures. See Figure 22:3.

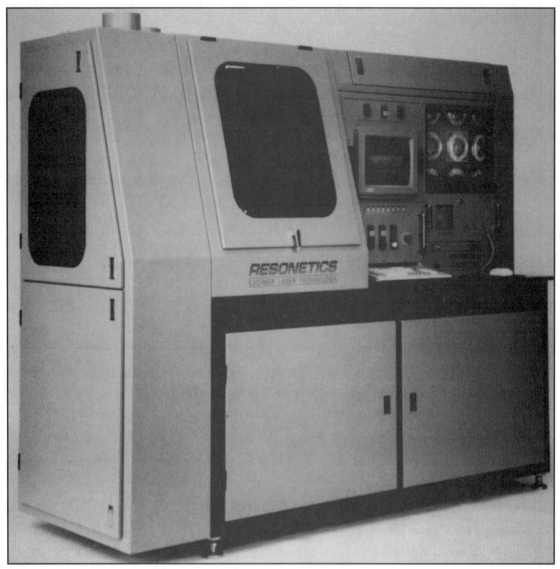

Courtesy Resonetics

Figure 22:3
A Micromachining Excimer Laser Center

How Excimer Lasers Work

Basically, excimer lasers do not use a thermal process. They ablate material by breaking down the molecular bonds of the material, which produces a plasma plume. The excimer laser, with its short wavelength, is essentially a heatless process, thereby producing virtually no heat-affected zones.

Capabilities of Excimer Lasers

By using a pulsed, deep ultra-violet excimer laser, hole diameters have been produced below 2 micron (.00008″). See Figure 22:4.

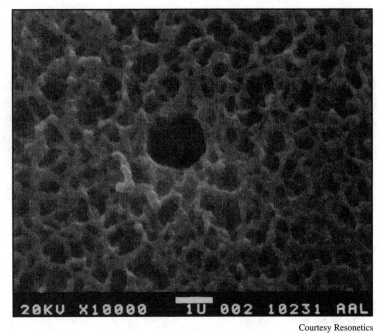

Figure 22:4
Scanning Electron Microscopic Photo of 1.5 Micron (.00006″) Diameter Hole

Excimer lasers use a system called masking. The laser pulses a beam through a mask containing the pattern to be machined. For example: To produce holes in .001″ (.025 mm) thick polyester, each hole had to be 25 microns (.001″) and on 50 micron (.002″) centers. With just one pulse from the excimer laser, 7,000 holes were produced, as illustrated in Figure 22:5.

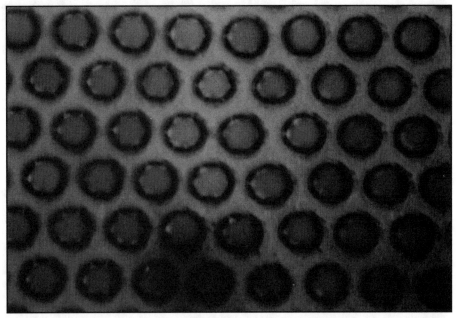

Figure 22:5
Micromachining: An excimer laser produced 7,000 holes with 25 micron (.001″) diameters on 50 micron (.002″) centers simultaneously on .001″ (.025 mm) thick polyester.

Because of its ultra-short wavelength, excimer lasers can produce miniature high image resolutions. See Figures 22:6 and 22:7

Courtesy Resonetics

Figure 22:6
A 40 micron (.0016″) reverse character image etched to a 40 micron depth on ceramic.

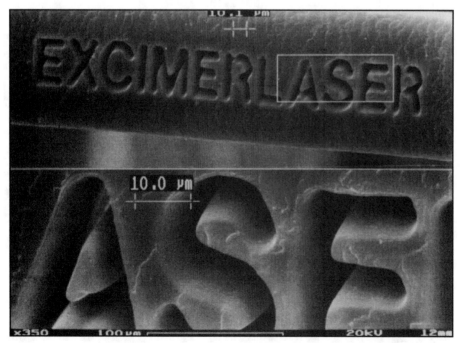

Courtesy Lambda Physik

Figure 22:7
Micro-machined letters on a single human hair.
Note the clarity of the letters in the close-up view.

Perhaps the greatest benefits of excimer lasers is their capability in the production of high-density printed circuit boards. Hybrid circuits for personal computers, mainframes, and super computers require the removal of various polyimide insulating materials and the creation of microscopic circuit patterns. See Figures 22:8 and 22:9.

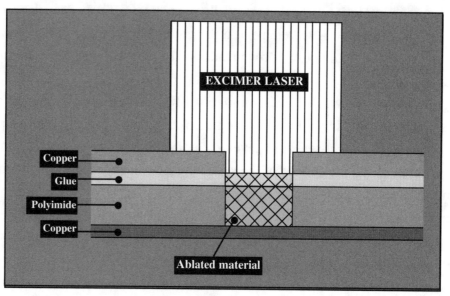

EXCIMER LASER

Copper
Glue
Polyimide
Copper

Ablated material

Figure 22:8
Illustrating the principle of selective polyimide ablation. The excimer laser does not damage the copper mask, but it removes the polyimide leaving a clean wall and base.

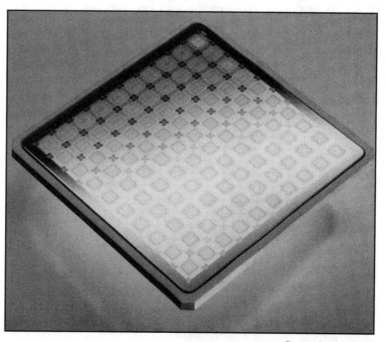

Figure 22:9
Glass Ceramic Substrate of a Multi-Chip Module

Optical Microlithography—Laser for Integrated Circuits

Integrated circuits (ICs), such as memory chips and microprocessors, can be found in computers, automobiles, planes, machines, phones—practically everywhere. Manufacturers are producing such small integrated circuits that millions of microscopic transistors have been placed on them.

The semiconductor industry is continually searching for ways to place more transistors on each chip. A key element in the production of these chips is optical microlithography. Continuous improvements are being made in lasers and optics to produce a high-quality, deep ultraviolet light.

Integrated circuits, which consist primarily of transistors, are produced on a single semiconductor. Layer by layer the microlithography process builds up the chip. Generally, a light-sensitive layer called a photoresist is coated on the semiconductor. Then the surface of the semiconductor is exposed through a mask by either a laser or lamp light. This process is similar to excimer laser. The difference is that with excimer laser, the material is ablated; whereas, with laser the light exposes the photoresist. This light exposure allows the features on the wafer to be significantly smaller.

Since the integrated circuits are so small, hundreds of these circuits can be produced from one wafer. The laser continually moves to another area and repeats the microlithography process. The device that performs this operation is called optical wafer steppers. When the wafer is totally exposed, a chemical is applied to the photoresist which produces the desired pattern. After this application another layer of photoresist is applied. The process can be repeated up to 20 times. See Figure 22:10.

The hindering factor in producing smaller chips is the wavelength of the light source. Manufacturers are continually searching for lasers to produce smaller bandwidths so integrated circuits can become more powerful for tomorrow's needs.

Courtesy Industrial Laser Solutions/Intel

Figure 22:10
Processor die contains more that 3 million transistors.

Traveling Methods of Laser Cutting Machines

Lasers have three basic traveling methods: beam traveling, workpiece traveling, and combination of beam and workpiece traveling. Each system has its advantages and disadvantages.

1. Beam Traveling

The work table remains stationary while the laser beam travels. This process is called flying optics. The advantage of flying optics is the weight of the material being cut has no effect on machine movement. This allows for rapid travel when the machine is cutting, and when the machine is travelling to another section.

The disadvantage is that the laser beam can vary slightly in diameter from one corner of the machine to another. To test for possible variance, the same diameter hole should be cut in each of the four corners of the machine and the difference measured.

2. Workpiece Traveling

The laser remains stationary while the workpiece travels. The advantage of this method is that the laser beam size remains constant. This method is also used when lasers are combined with turret punch presses.

The disadvantage of the workpiece travelling method is that the weight of the workpiece can affect cut accuracy. To produce accurate cuts, slower table speeds are used.

3. Combination of Beam and Workpiece Traveling

A hybrid system is commonly used where the laser moves in the X direction and the table moves in the Y direction. The advantage is that there is a smaller variation in the beam path length. This stability produces greater accuracy than the beam traveling method.

The disadvantage is that it is not as fast in high-speed machining as the beam traveling method.

Custom-Made Laser Systems

Custom-made laser systems can be built for specific operations, or a basic laser system can be purchased and various beam delivery systems can be used, as shown in Figure 22:11.

A. **Series BSH-1500 Beam Shutter** provides a physical barrier which blocks and dumps the laser beam during beam interruption.

B. **Series BSP-45 Beam Splitter** for splitting into multiple power beams.

C. **HeNe Laser Injection Port**

D. **Series PVR-6000 Polarization Vector Rotator** for rotation of the linear polarized vector of any high power CO_2 laser beam to any desired angle while preserving the linear polarization.

E. **Series RP-3000 Reflective Polarizer** with adapters to isolate back-reflection from the work piece before it hits the laser.

F. **Series SWVL Rotation Swivel** to be used in conjunction with any BD&D device that requires manual or automatic continuous rotation.

G. **Series LCP-6000 Linear to Circular Polarization Converter** for the conversion of linear polarization to circular polarization.

H. **Series BBN-90 Beam Bender** to redirect the laser beam, at 90° from the entering beam.

I. **Series TBNG Beam Enclosure Tubing**

J. **Series BSL-3000 Beam Shuttle** for air-pressure-operated beam switching to redirect the beam energy.

K. **Series SFL-500 Spatial Filter** used to improve the beam quality by reducing the spatial noise.

L. **Series ADP Adapter Spacer**

M. **Series COL-1500 Expander/Collimator** that input a collimated or diverging/converging beam and output an expanded collimated beam.

N. **Series ACH-101 AccuTRACK™ Surface Contact Cutting Head** for cutting of sheet metal up to 1/4″ thick, non-metallic material cutting, light duty welding.

O. **Series ACH-145 AccuTRACK™ Surface Contact Cutting Head** for cutting of sheet metal up to 1/4″ thick.

P. **Series AWH-126 AccuWELD™ Parabolic Mirror Welding Head** for welding and cutting of steel and other metals at powers up to 20kW.

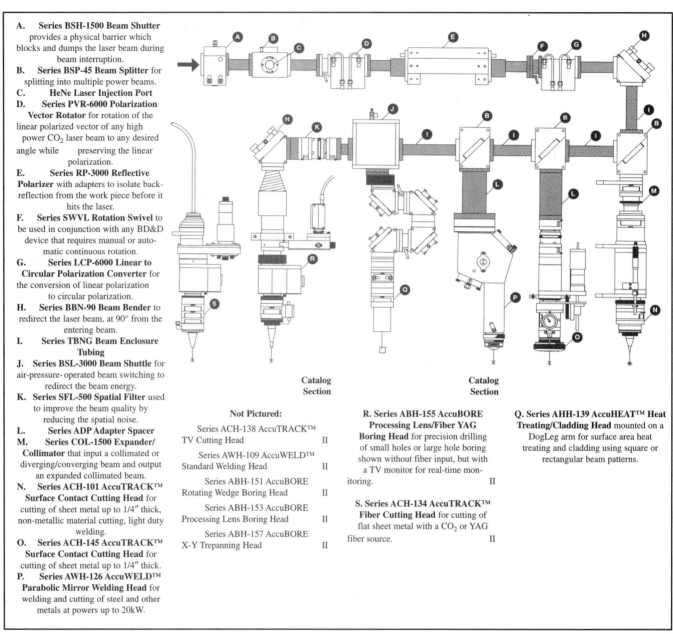

Not Pictured:

Series ACH-138 AccuTRACK™ TV Cutting Head   II

Series AWH-109 AccuWELD™ Standard Welding Head   II

Series ABH-151 AccuBORE Rotating Wedge Boring Head   II

Series ABH-153 AccuBORE Processing Lens Boring Head   II

Series ABH-157 AccuBORE X-Y Trepanning Head   II

Catalog Section

R. **Series ABH-155 AccuBORE Processing Lens/Fiber YAG Boring Head** for precision drilling of small holes or large hole boring shown without fiber input, but with a TV monitor for real-time monitoring.   II

S. **Series ACH-134 AccuTRACK™ Fiber Cutting Head** for cutting of flat sheet metal with a CO_2 or YAG fiber source.   II

Catalog Section

Q. **Series AHH-139 AccuHEAT™ Heat Treating/Cladding Head** mounted on a DogLeg arm for surface area heat treating and cladding using square or rectangular beam patterns.

Courtesy Laser Power

Figure 22:11
Various Beam Delivery Systems

Beam Splitting

Laser beams can use electrically controlled optical devices that split the beam into two or more separate beams. Therefore, several locations can use one power source.

A three station beam-splitting laser allows all stations to receive the same power. The first-beam splitter should reflect 33.3% and allow 66.6% of the power to pass through. The second splitter should reflect 50% of the power and allow 50% to go through. The third splitter should reflect all the power received.

Now that various lasers have been analyzed, the next section will discuss the many ways of profiting with laser cutting.

23 Profiting With Laser Cutting

Materials for Laser Cutting

An understanding of the capabilities and advantages of lasers helps manufacturers utilize their many benefits. Listed below are some of the materials that can be cut with lasers.

Superalloys	Titanium	Plastic
Stainless Steel	Aluminum	Kevlar
Tool Steels	Brass	Wood
Carbon Steels	Copper	Bakelite
Galvanized Steel	Monel	Fiber
Hardened Steel	Beryllium Copper	Paper
Inconel	Ceramics	Masonite
Hastalloy	Composites	Leather
CPM 10V	Screen Materials	Cork
Carbide	Gasket Materials	Fused Silica

Determining Factors on Material Thickness

Any material, whether metallic or non-metallic, can be cut with lasers. There are two considerations in cutting various materials—the optical and the thermal characteristics of the material. The optical is the ability of the material to reflect the laser wavelength, whereas the thermal is the ability of the material to absorb the laser wavelength.

Since lasers cut by means of high energy light, some difficulties are encountered with highly reflective materials such as gold and silver. For these materials, mechanical cutting methods are generally more cost effective. Also, the high thermal conductivity of aluminum and copper causes laser beam heat to be dissipated quickly. As a result, a 1500 watt CO_2 laser cuts only up to 1/8″ (3.2 mm) aluminum and up to 1/16″ (1.6 mm) copper. Laser wattage determines how thick a laser can cut. A 1500 CO_2 laser can cut carbon steel up to 1/2″ (13 mm). A 3000 CO_2 watt laser can cut carbon steel up to 3/4″ (19 mm).

Listed below are other material thicknesses that can be cut with a 1500 watt CO_2 laser:

High Alloy Steels—.001 to 1/2″ (.025 to 13 mm)
Stainless Steels—.001 to 3/16″ (.025 to 4.8 mm)
Plexiglass—to 1″ (25 mm)
Pressboard—to 3/16″ (4.8 mm)
Rubber—to 3/16″ (4.8 mm)
Wood, Glued—to 5/8″ (16 mm)
Wood, Oak—to 1″ (25 mm)

Lacquered Metals

The ability to cut lacquered metals with lasers varies substantially according to the type of surface paint used. One solution to cutting lacquered metals is to first use an engraving cut to remove the lacquer, then repeat the program as a cutting operation.

Tube Cutting

Some lasers are equipped to cut tubes. The "Z" axis is programmable with a rotational axis. These three-axis lasers can cut various types of tubing, as pictured in Figures 23:1 and 23:2.

Courtesy Bystronic Laser

Figure 23:1
Lasers Cutting Tubes

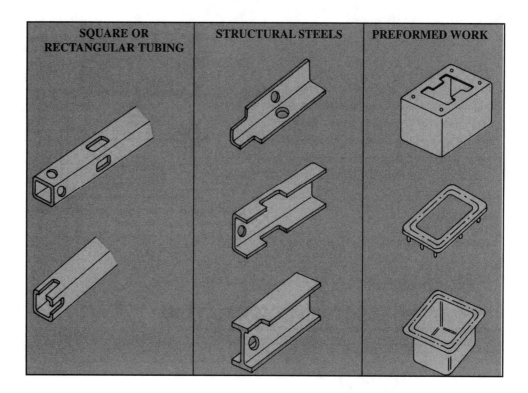

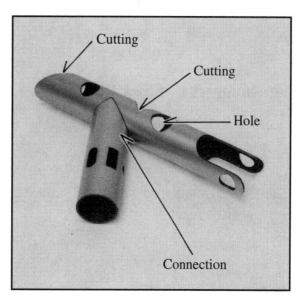

Figure 23:2
Various Tube and Preformed Cutting Applications

Multiaxis Laser Cutting

There are multiaxis lasers that have five and six axes (X, Y, Z, A and B for five axis lasers, and a W axis for the six axis lasers). The sixth axis, or the zero-offset head, provides a faster response time, particularly when the laser needs to cut around corners. Multiaxis lasers can cut various shaped materials, as shown in Figure 23.3.

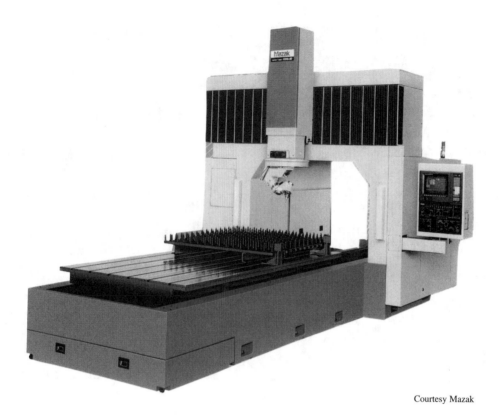

Courtesy Mazak

Figure 23:3
A Six Axis Laser

Lasers With Pallet Changers or Sheet Loaders

To increase productivity, some lasers come equipped with pallet changers or sheet loaders. Instead of the laser being idle while the operator removes the cut parts, an unloader system removes the cut sheet and replaces it with one to be processed. While the laser is cutting, the operator removes the parts from the previous sheet. See Figures 23.4 and 23.5.

Courtesy Mitsubishi Laser

Figure 23:4
Automatic Pallet Changer

Courtesy Bystronic Laser

Figure 23:5
Automatic Shuttle Table With Unloader

Lasers With Robots

Lasers can be attached to conventional robots. Nd:YAG lasers can be used with fiber optics to produce an efficient and cost effective laser operation

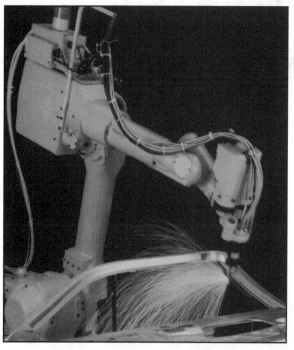

Courtesy Fanuc Robotics

Figure 23:6
A Multiaxis Nd:YAG Laser With Fiber Optics

Part Trimming

Often when parts are formed, the edges need to be trimmed. Multiaxis lasers are ideal for these trimming procedures, as pictured in Figure 23:7.

Courtesy Lumonics

Figure 23:7
A Multiaxis Laser Doing Part Trimming

Time Sharing

Fiber optics allow one laser to perform various production tasks. Some lasers are able to switch up to 40 times per second between fibers. Eight fibers can be attached to one laser, and a controller between the laser and switching unit can alter the laser parameters to allow for either drilling, cutting, or welding at each station. A time share application is depicted in Figure 23:8.

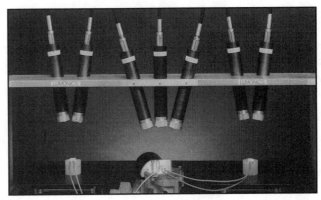

Courtesy Lumonics

Figure 23:8
Time-Share System for On-Line Soldering Applications

Advantages of Laser Cutting

1. Material Savings

Laser cutting creates a very narrow kerf, from .006 to .010″ (.15 to .25 mm) wide. These narrow kerfs results in substantial material savings.

2. Minimum Heat-Affected Zone

Since most of the thermal energy is absorbed by the molten material in the kerf, lasers have a small heat-affected zone.

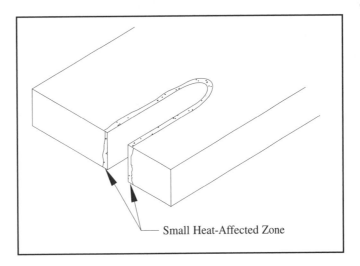

Small Heat-Affected Zone

Figure 23:9
Minimal Heat-Affected Zone

3. Edge Quality

Lasers produce fine finishes on thin pieces, but the edge deteriorates as the cut gets thicker. For most applications, however, even on thick materials the edge quality is acceptable.

4. Minimum Distortion and Thermal Stress

The limited heat-affected zone minimizes material distortion and thermal stress.

5. Close Nesting of Parts

Due to the narrow kerf and low-stress machining of laser cutting, parts can be closely nested. See Figure 23:10.

Figure 23:10
Close Nesting of Parts

6. Thin Webs

The absence of mechanical forces, as in stamping and turret presses, allows lasers to cut extremely thin webs without distortion, as shown in Figure 23:11

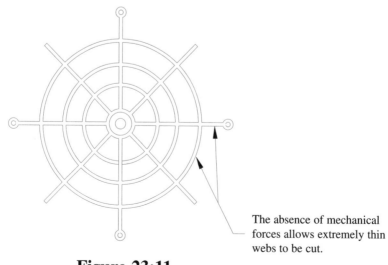

The absence of mechanical forces allows extremely thin webs to be cut.

Figure 23:11
Thin Web Cutting

7. Elimination of Hard Tooling

Building hard tooling is expensive, whereas with lasers, a programmed part can be quickly produced. Laser cutting also eliminates the need to store various kinds of tooling.

8. Just-In-Time (JIT) Machining

Just-in-time delivery is greatly enhanced because the need for making special tooling is eliminated. Parts can be made quickly from the drawing board to reality.

9. Various Shapes and Sizes

Designers should realize that cutting complicated shapes can be produced easily with lasers. When thinking of hard tooling, designers must consider its limitations; but whatever can be drawn on a computer can be cut with a laser. See Figures 23:12 and 23:13.

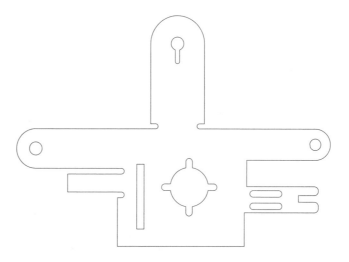

Figure 23:12
Any Imaginable Shape Can be Economically Produced

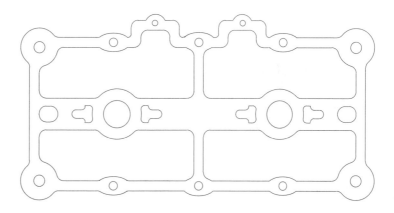

Figure 23:13
Laser Cut Gasket

10. Material Hardness

Unlike conventional tooling, lasers cut material of any hardness.

11. Consistent Laser Beam

While conventional cutters wear and become dull, properly focused laser beams remain consistent.

12. Prototypes Can be Fabricated Easily

Prototypes can be quickly built, as in Figure 23:14. If alterations are needed, a new prototype can be easily programmed and reproduced.

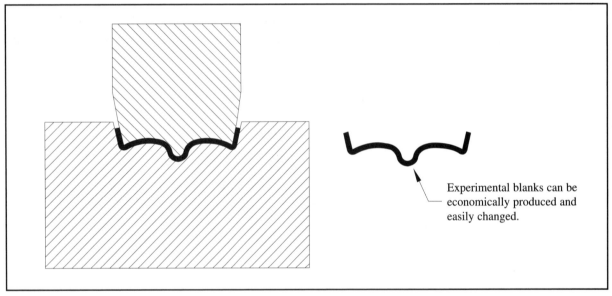

Experimental blanks can be economically produced and easily changed.

Figure 23:14
Prototypes Can be Easily Fabricated

13. Repeatability

By storing programs on the computer or disk, operators can reproduce an exact duplicate repeatedly if all the cutting parameters stay the same.

14. Ideal for Short Run Production

Making short runs with conventional tooling can be very expensive. Because lasers do not need any special tooling, they produce short runs quickly and efficiently.

Disadvantages of Lasers

One of the great disadvantages of lasers is their high cost. There must be sufficient volume of work to justify purchasing a laser. Also, many lasers require substantial maintenance. Gases created during the laser process can cause splatter which can damage the sensitive optics. Also the slightest impurities in the laser gases can cause the internal mirrors in the resonator to become ineffective.

Industrial Laser Review reported the results from representatives of laser job shops who asked a panel of laser systems manufacturers questions about laser cutting problems. A job shop owner asked about reducing the cost of operating a laser: "Our shop runs two systems seven days/week, 24 hours/day, and we currently spend $4000-$5000 per machine per month for total maintenance."

A laser company representative replied, "Laser optics remain the predominant maintenance item, and they run at about $4/hr. When you start multiplying that number ($4 x 22.5 average days per month x 24 hrs/three shifts) it does add up ($2160/month). We have found that the quality and cleanliness of the laser gas can significantly impact the life of laser optics."

In spite of some of the disadvantages of lasers, lasers have proven to be an efficient and cost effective method for machining parts for many manufacturers. Other costs and issues to be considered when purchasing a laser system will be discussed in Chapter 27, "Purchasing the Right Equipment."

24 Lasers for Welding, Cladding, Alloying, Heat Treating, Marking, and Drilling

I. Laser Welding

The two common lasers used for welding are Nd:YAG and CO_2. Welding occurs when a high intensity focused laser beam is directed and joins two or more similar or dissimilar metals by melting them together. See Figure 24:1. An inert gas of argon or helium generally surrounds the laser beam to protect the weld from oxidation.

Courtesy Lumonics

Figure 24:1
Parts Being Laser Welded

The focused laser beam allows the fabrication of a very narrow weld, and produces a minimum amount of heating and thermal distortion. Easily distorted and delicate parts are good prospects for laser welding because of the low heat input. Foils as thin as .0005″ (.0127 mm) have been welded.

A. Laser Welding Compared to Electron Beam Welding

Laser welding derives its energy from photons; with electron beam welding, the energy comes from electrons. The electron energy collides with the solid material causing thermal energy. When sufficient energy is applied, the electrons hitting the surface melt the material.

Electron beam welding is done in a vacuum chamber. Welding in a vacuum chamber allows for a wide range of control and insures purity in the welding process. Laser welding has the advantage of being performed in the open atmosphere.

B. Two Types of Laser Welding: Spot and Continuous Welding

Spot Welding: two overlapping metals are joined together by means of a laser beam melting an area that penetrates both metals (Figure 24:2).

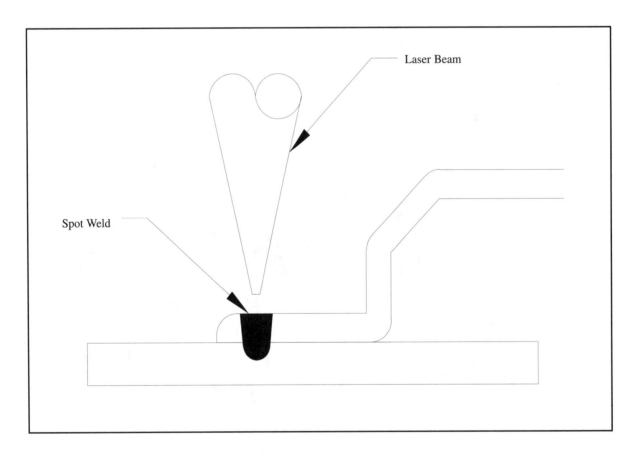

Figure 24:2
Intermittent Laser Spot Welding

Continuous Welding: a laser beam continuously moves along the surface of two metals and melts both together to form a permanent continuous union. Various joint designs can be welded in this manner, as shown in Figure 24:3 and 24:4.

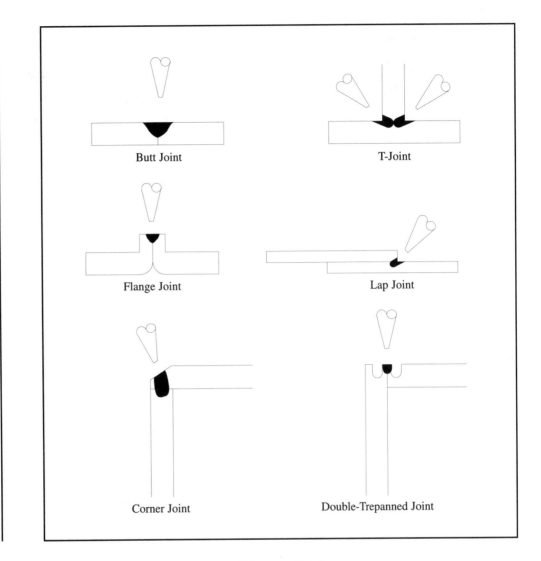

Figure 24:3
Various Laser Joint Designs

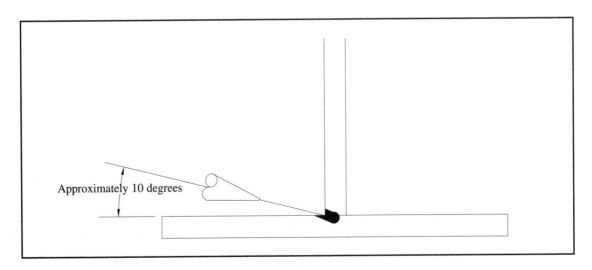

Figure 24:4
Skid Welding: A Single-Pass Welding for T-Joint Welds
The drawing shows a way to weld just one side of a joint. The process is
known as skid-welding. The laser nozzle is aimed at the joint at about 10 degrees.

C. Welding Dissimilar Metals

Continuous-wave butt welding can weld dissimilar metals. Welding a copper alloy to a steel alloy, as shown in Figure 24:5 , is performed by focusing the laser beam on the steel alloy so that both metals are joined. The reason for aiming the laser beam on the steel alloy is due to high reflectivity of copper.

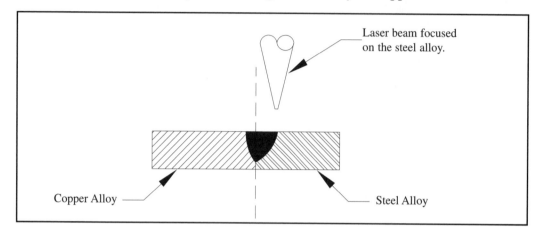

Figure 24:5
Welding Dissimilar Metals

D. Keyhole Welding

Lasers are capable of performing deep welds known as keyhole welds. The welds can have a depth to width ratio of over 10:1.

Keyhole welding occurs when a high power laser beam penetrates the metal and causes it to melt. The center of the weld begins to boil, while the outer edges of the molten material cling to the solid material. The surrounding cooler edges where the weld is being done have a higher surface tension; therefore the cooler edges pull the molten material to form a narrow pocket. Vapor pressure from the boiling material, as illustrated in Figure 24:6, prevents the narrow pocket from collapsing.

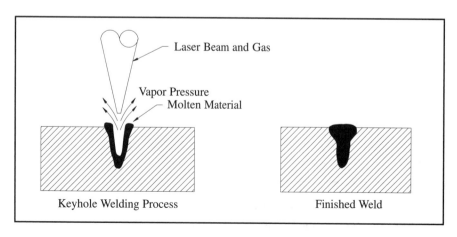

Figure 24:6
Keyhole Welding

E. Focusing the Beam and Through-The-Lens TV Viewer

Some lasers come equipped with cross hair binoculars in order to focus the beam. Other lasers have a through-the-lens TV viewer. The part can be programmed and welded, drilled, or cut from viewing the monitor.

Although the weld plume affects capacitance sensors, some lasers are equipped with automatic focus control that can focus in spite of the weld plume.

F. Considerations for Laser Welding

Many applications exist for laser welding. However, the initial high cost of lasers presents needs to be considered when comparing the price of conventional welding equipment.

In conventional spot welding, the spot welder makes physical contact with the metal and forces the metal together. With laser spot welding, the laser never touches the surface. When spot welding with lasers, a clamping device should be used to guarantee a proper laser spot weld.

When laser welding is performed by CNC methods, see Figure 24:7, there must be sufficient quantity of work to justify CNC laser welding compared to TIG, MIG, or other much less costly welding procedures.

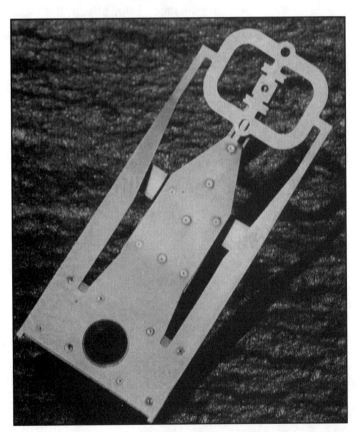

Courtesy Lumonics

Figure 24:7
Sixteen spot welds done in two seconds for a magnetic head of a computer disc drive.

Most laser welding receives no filler metal. Therefore surfaces to be laser welded must be aligned accurately to assure proper welding. Although laser welding is costly and may require fixturing, many manufacturers have proven that laser welding is economical due to its:

- accuracy and consistent welds
- minimal workpiece distortion
- no filler or flux required
- automating possibilities
- high speed

- low heat-affected zone
- narrow weld head
- welding of dissimilar materials
- ability to perform either spot or continuous welding

II. Laser Cladding

Laser cladding adds material to a surface so it can withstand greater wear. The advantage of cladding is that an inexpensive material can be coated with an expensive alloy, such as stellite. In addition, the soft undercoat of the hardface makes the part more durable.

Laser cladding is performed when a laser beam is focused on a desired surface which has been prepared with a powder substrate, or while a powdered alloy for hardfacing is applied, as depicted in Figure 24:8. The laser melts the applied alloy, and the alloy adheres to the underlying surface. Four different types of material are generally used for cladding: iron base, nickel base, cobalt base, and carbide.

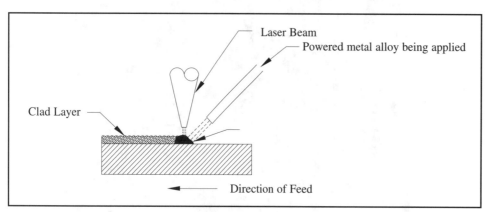

Figure 24:8
Laser Cladding

A process similar to laser cladding is laser surface melting. A thin layer of material is placed on the workpiece for the laser to melt. As the surface material melts and is 'self-quenched,' it creates beneficial surface properties on the workpiece.

A. Conventional Cladding

Some conventional cladding methods are flame spray, plasma spray, thermal spraying, metallizing, weld overlay, MIG, TIG, fuse weld, and plasma arc. Some of the systems used for cladding are high-velocity oxygen fuel spray, plasma spray, two-wire electric arc spray, and TIG weld overlay.

Cladding methods such as TIG weld overlay cause difficulties because the applied hard face is usually uneven and requires extensive grinding. Other problems occur, such as the overuse of expensive hardfacing materials, and the applied high heat can cause alloy dilution and part distortion.

B. Advantages of Laser Cladding

Cladding with lasers, however, allows the hard face thickness to be closely maintained. Speed, laser power, and the powder feed can be controlled precisely, to produce a continuously evenly-laid hard face. Some laser hardfacing requires only .005″ to .010″ (.13 mm to .25 mm) of machining to obtain the desired smoothness.

III. Laser Surface Alloying

In laser surface alloying, the metal surface is brought to the melting point as an applied material mixes chemically with the surface. The added material not only fuses to the surface, but it also thoroughly mixes with the surface, thereby creating a new alloy after solidification.

A. Lasers the Ideal Tool for Surface Alloying

To achieve the desired alloying of the surface, there must be exact controls. Lasers are an ideal tool to accomplish this because of their ability to control accurately the amount of energy.

B. Difference Between Laser Alloying and Laser Cladding

The difference between surface alloying and cladding is that cladding lays new metal on the surface. Alloying introduces metal onto a heated surface, and this new metal mixes with the surface and creates a new alloy. See Figure 24:9

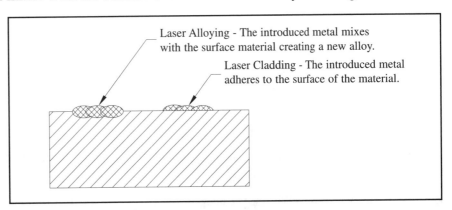

Figure 24:9
The Difference Between Laser Alloying and Laser Cladding

IV. Laser Heat Treating

Laser heat treating, or what is more precisely described as surface transformation hardening, is widely used to improve the surfaces of machined parts.

A. Procedures for Laser Heat Treating

Generally, laser beams emit an intense, narrowly focused beam of light. For heat treating, however, the laser beam is defocused in order to cover a larger area. The shape and size of the laser beam is achieved by using special optical systems designed for this purpose. The round beam can resemble a doughnut, or it can become a square beam altered by an optical integrator, as shown in Figure 24:10.

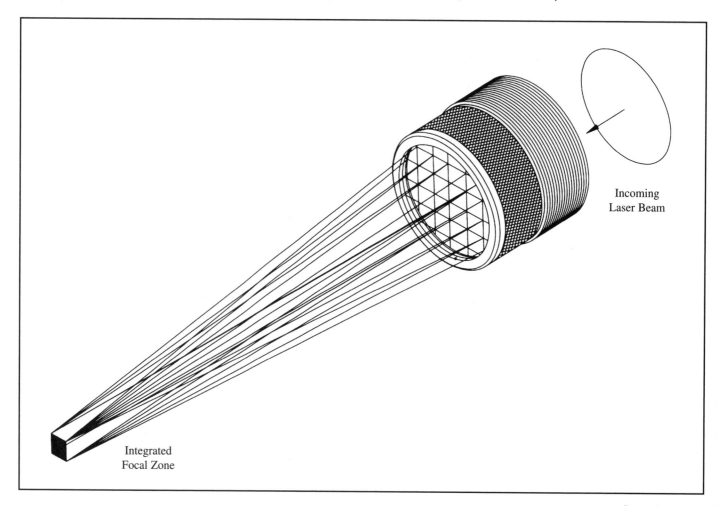

Incoming
Laser Beam

Integrated
Focal Zone

Courtesy Laser Power

Figure 24:10
A Round Beam Converted to a Square Beam

In laser heat treating, the laser beam heats the workpiece surface to above the austenitic temperature, but below melting. Since the surface is only heated, the cool interior of the workpiece quickly dissipates the heat when the beam passes. Quenching and hardening occurs as the workpiece cools.

Laser heat treating is suited primarily for localized hardening and on self-quenching metals. The depth of the case hardening and its hardness are determined by the workpiece material, the amount of applied laser energy, and the processing speed. See Figure 24:11.

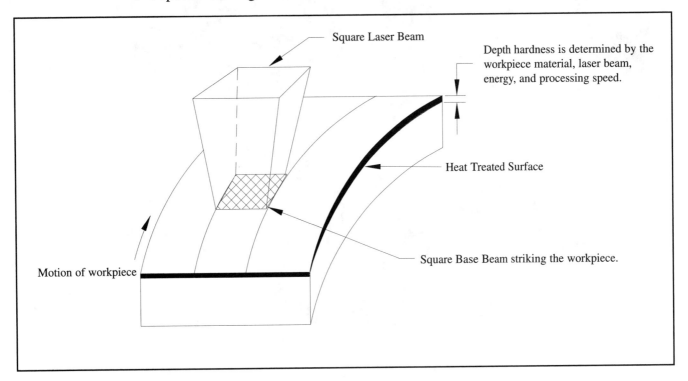

Square Laser Beam

Depth hardness is determined by the workpiece material, laser beam, energy, and processing speed.

Heat Treated Surface

Square Base Beam striking the workpiece.

Motion of workpiece

Figure 24:11
Laser Heat Treating

B. Advantages and Disadvantages of Surface Heat Treating

Since conventional heat treating heats the entire part, it can cause distortion and result in having a part that is too hard to withstand shock. Surface heat treating hardens only the outside of the workpiece while the internal area remains soft. The advantage of surface heat treating is it increases the wear resistance of the workpiece, and allows it to withstand greater shock.

The disadvantage of laser heat treating is the high cost of the laser equipment, and the expense involved in individually heat treating each part compared to conventional batch heat treating.

V. Laser Marking

Lasers are used to mark hardened bearing cages, rulers, carbide drill bits, circuit boards, surgical instruments, labels requiring serializing, and many other products. Lasers can engrave all sorts of metals, plastics, ceramics, and other materials on both flat and round surfaces. See Figure 24:12.

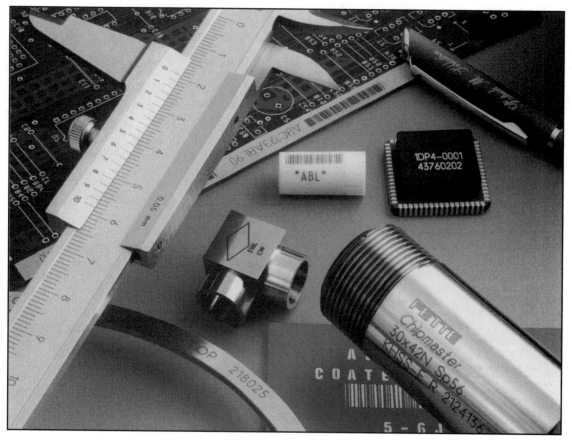

Figure 24:12
Various Laser-Marked Parts

A. How Laser Marking Works

A narrowly focused laser beam can melt, anneal, or engrave by vaporizing the surface. This action changes the surface, as illustrated in Figure 24:13.

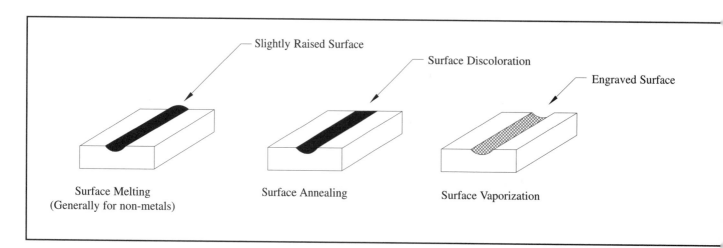

Figure 24:13
Three Methods of Surface Alterations for Laser Marking

When lasers are focused on anodized metal or coated material, as shown in Figure 24:14, the beam removes the top coated dark layer and exposes the highly visible white second layer.

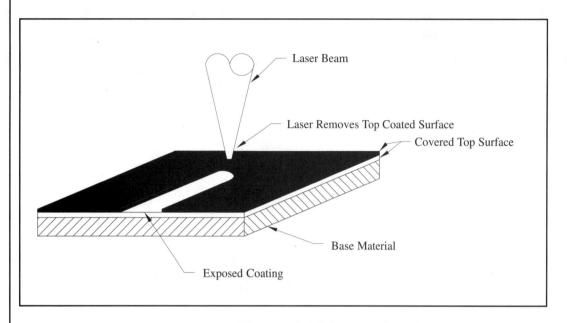

Figure 24:14
Laser Marking on Coated Surfaces

Laser marking devices rapidly switch peak power on and off. This allows swift movement to another position to continue its marking. Lasers can do either dot matrix marking or continuous line marking. Some marking lasers use a mask or a stencil to engrave; others use a beam-steered image method.

The mask laser is a speedier process than the beam-steered method which must draw the image. However, the beam-steered method can create a better image and is more flexible. The beam-steered laser can mark any computer generated image, and it can quickly alter the image for a new part.

B. Advantages of Laser Marking

A major advantage of laser marking is the speed in which laser can engrave parts. Coupled with automation, parts can be marked rapidly. In one automated procedure, a photoelectric cell triggers a marking laser that has been activated by the part on a conveyor belt.

One company coded their capacitors with an ink pad technique at a rate of 60 parts a minute, but when they installed laser marking, the procedure increased to 300 parts a minute. The speed for marking was limited by the associated test equipment, not the laser. The ink pad procedure occasionally created illegible marks; the laser increased the quality by making clear and permanent marks.

Chemical etching can leave an acid residue which can produce undesirable effects. In contrast to chemical etching and ink or dye printing, lasers do not contaminate the surface. This ability for lasers to do non-contaminated marking is critical for medical equipment. Laser marking is also permanent because the surface has been altered. Even the hardest substances can be marked with lasers.

Other marking procedures often require complex holding fixtures. Because laser marking is a non-contact procedure, lasers need no fixtures.

VI. Laser Drilling

Laser drilling performs close-tolerance high-speed drilling in practically all materials, from super alloys to soft rubber. Some lasers can drill up to 50 holes a second. See Figure 24:16 for a laser drilled gas turbine.

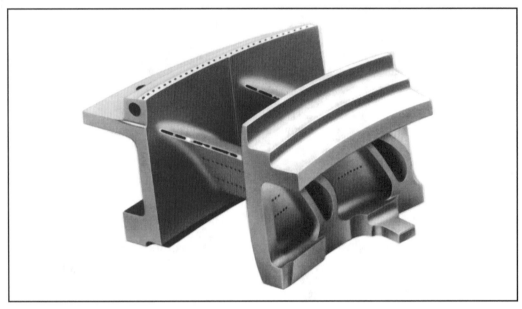

Courtesy Lumonics

Figure 24:16
Laser Drilled Gas Turbine

Lasers also drill materials as hard as ceramic. Generally laser drilled holes range from .0005″ to .060″ (.013 mm to 1.5 mm) to a depth of .600″ (15 mm). Some lasers can drill a .020″ (.5 mm) hole up to 1 1/2″ (38 mm) deep. Laser can also drill at low angles. Some lasers have been able to drill blind holes to controlled depths up to .32″ (8.1 mm) deep.

A. How Laser Drilling Works

A laser beam is produced with a lens that creates a long focal length. The pulsed laser energy heats, melts, and vaporizes the material. Pulse depth is determined by the material, beam size, pulse energy, pulse length, and peak power.

When starting the hole with lasers, additional pulses extend its depth. The laser beam is then reflected by the side walls of the hole causing a "light pipe effect," as illustrated in Figure 24:17.

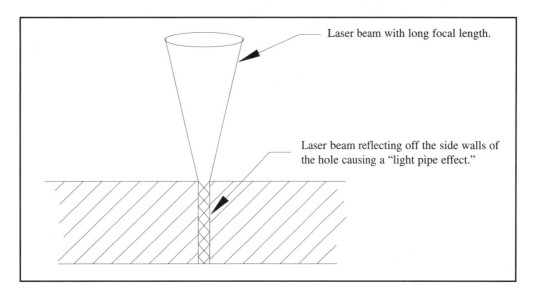

Laser beam with long focal length.

Laser beam reflecting off the side walls of the hole causing a "light pipe effect."

Figure 24:17
Laser Beam for Drilling

A small amount of melted material resolidifies and creates a recast layer on the sides of the hole. The thickness of this recast layer is determined by the laser energy, material being drilled, and assist gas parameters. Sometimes microcracking can occur.

For larger holes, a method called trepanning is used. The focused laser beam moves over the workpiece or the part moves. This method technically performs laser cutting. With trepanning, square or shaped holes can be produced.

When drilling, the laser lens must be protected from the melted material. A gas-assist nozzle is often used to protect the lens. Compressed air or oxygen can be used, this also aids in the cutting process. When non-metals are drilled, an inert gas, such as argon or nitrogen, is sometimes applied.

Surfaces adjacent to the exit of the drilled holes need to be protected from the laser beam and its splatter. To avoid splatter from the exiting laser beam, sometimes wax, teflon, or some other backup protection material is applied.

B. Disadvantages of Laser Drilling

- High cost of laser equipment
- Recast layer
- Entrance/exit ratio of hole diameters (typical taper is 1% of drilled depth)
- Slower processing for large holes

C. Advantages of Laser Drilling

- High-speed drilling
- Drills hardened materials
- Reaches difficult locations
- Produces accurately small to microscopic holes
- Eliminates drill breakage or sharpening
- Drills on curved surfaces
- Produces holes at difficult entrance angles
- Minimum heat distortion
- No tooling costs or wear
- Easy automation
- No chips
- Practically zero machining forces
- Rapid changing of hole diameters and shapes

The Future of Lasers

Many exciting developments are happening with lasers. These developments include through-the-lens cameras, sensors unaffected by welding plume, beam analysis tools, ultra-violet sensors, reference point locators, automatic orientation of the beam director, built-in cameras teaching operators how to program 3D parts, adjustable splitters, higher power lasers, and digital resonators.

Modern lasers are showing a remarkable increase in cutting speed. Within a ten year period, one laser manufacturer reports that cutting a .400″ (10mm) hole to an accuracy of .002″ (.05 mm) has gone from 55 inches per minute (1400 mm) to 320 inches per minute (8120 mm). In one minute, the number of holes produced has increased from 11 to 145, an increase of 1300 percent.

As has been shown, dramatic improvements have been made in lasers, and they are extremely useful for many applications. The challenge lies ahead for manufacturers, designers, and engineers to learn and use the laser's unique capabilities to increase productivity.

Unit 11

Rapid Prototyping and Manufacturing

Notes

25 Rapid Prototyping and Manufacturing

Rapid Prototyping and Manufacturing, also known as Three-Dimensional Printing or Virtual Prototyping and Manufacturing, is one of the newest, most exciting developments in parts and molds production made possible through the technology of virtual reality. See Figures 25:1 and 25:2. Virtual reality describes an artificial environment in which any of the human senses—sight, sound, taste, touch, or smell—can be simulated. Virtual reality is simply a computerized simulation of actual experiences encountered in the real world. Today, dramatic changes are taking place in manufacturing because of rapid prototyping.

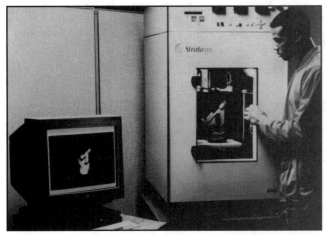

Courtesy Stratasys

Figure 25:1
Rapid Prototyping: From Computer to Model

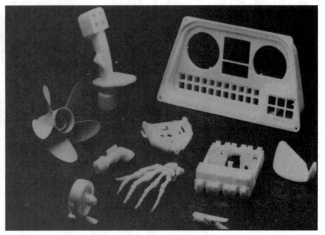

Courtesy Stratasys

Figure 25:2
Parts Made With Rapid Prototyping

In the past, design engineers would have designed parts produced elsewhere, and then have them machined or sculptured from wood, plastic, metal, or other solid material. Now with rapid prototyping, solid parts can be manufactured overnight directly from a computer program that previously took days, weeks, and sometimes even months to produce. Engineers are now able to check their computer drawings with physical models and make necessary changes. Parts are also now being produced, not just prototypes, but functional manufacturing tools with rapid prototyping.

Today, some of these three-dimensional prototype machines (also called desktop modelers, design verification tools, and concept modelers) are as quiet as desktop printers. They are now appearing in offices of design engineers. These desktop modelers resemble a copier, fax, or printer rather than a tooling machine.

These desktop modelers allow engineers and manufacturers to evaluate in the early stages of the design any shortcomings from the computer design. Often the biggest risk is in the early developments of the design where knowledge about the part is limited. Having a 3D model to evaluate the functionality of the object in these early stages, and having the ability to make easily any necessary changes, is a great asset for bringing a quality part to the market quickly and at low costs. Also rapid prototyping can be used for photo-optic stress analysis and dynamic vibrational analysis. See Figure 25:3.

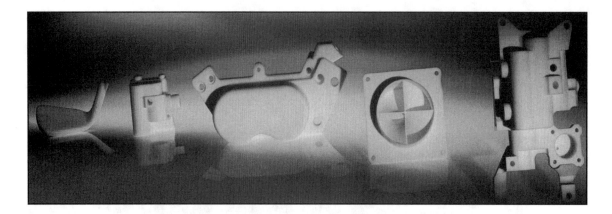

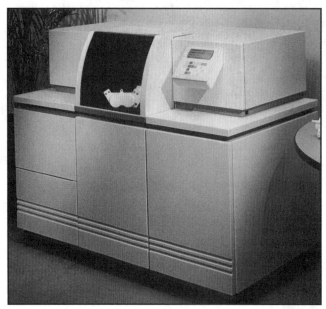

Courtesy 3D Systems

Figure 25:3
Desktop Modeler System and Three-Dimensional Models

In 1986, a patent was issued for rapid prototyping. Various systems have since been developed, and now these processes are being used in manufacturing plants around the world. These processes use CAE/CAD/CAM(computer-aided engineering/computer-aided design/computer-aided manufacturing) to create the desired part without the use of conventional machines.

The ability of computers to slice three-dimensional drawings has produced various systems that can create replicas of the computer images. This unique process of making parts directly from computers, leads to many exciting and profitable possibilities. The following section briefly describes some of the interesting rapid prototyping systems.

The Photopolymer Rapid Prototyping Process

A. Computer Aided Design

The first step in the rapid prototyping process is to create the desired three-dimensional computer aided design (CAD). The design model is then converted to a rapid prototyping format. The part is then sliced into horizontal planes from .0025″ to .030″ (.063 mm to .76 mm) thick. The usual slices are from .004″ to .008″ (.10 mm to .20 mm).

B. Laser Drawing

Then a Helium Cadmium (HeCd) laser beam controlled by the CAD program quickly passes over the liquid photopolymer resin. When the ultraviolet (UV) photons strike the photosensitive resin, it hardens. The computer calculates the exact time needed for the laser beam to harden the resin, and the optical scanning device moves the laser precisely according to the programmed path.

C. Photopolymerization

Stereolithography utilizes a process known as photopolymerization—a liquid plastic monomer is changed into a solid polymer as it is exposed to a laser beam. The degree of hardness of the photopolymer is determined by the amount of energy it receives. In stereolithography, the laser beam only hardens the polymer that it strikes; the remaining resin remains liquid and can be reused.

D. Translator

The laser beam does not strike the entire surface of the resin; rather, it rapidly draws the part border, and then performs what is know as "hatching." Hatching is accomplished by taking the CAD program and having it go through a rapid prototyping translator. This translator takes the desired shape and creates many small triangles or squares. Later on the part goes through a curing process.

E. Sweep and Z-Wait

To make certain that the resin is evenly distributed, a recoater blade sweeps rapidly over the previously hardened surface.

The laser beam travels again over the surface. As it moves over the resin, the newly hardened surface adheres to the one below it. This hardening and lowering of the Z axis is repeated until the desired shape is created. See Figures 25:4-5.

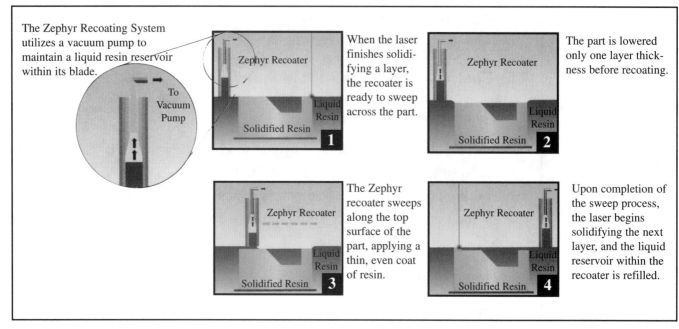

Courtesy 3D Systems

Figure 25:4
A Photopolymer Rapid Prototyping System

Courtesy 3D Systems

Figure 25:5
A Completed Part

F. Postcuring

The part then goes for final curing. The resin is in a "green state"—it has been only partially polymerized. A high intensity broadband or continuum ultraviolet (UV) radiation is used to cure the mold. Figure 25:6 shows a distributor cap that was made with the photopolymer rapid prototyping process.

Courtesy 3D Systems

Figure 25:6
Distributor Cap Made With The Photopolymer Rapid Prototyping Process

G. Finishing

The part at this stage is often satisfactory. However, to get a better surface finish, hand-sanding, mild glass beading, polishing, painting, or spray metal coating can be used. When either acrylate or epoxy based resins are used to make the object, the object can then be milled, bored, drilled, or tapped. Most parts are made with epoxy-based resins because they are much more accurate than parts made from acrylate resins.

H. Size Capabilities

Large parts can be made beyond the envelope of the vat. The part is produced in sections, and then the sections are welded together by using photopolymer.

Laminated Object Prototyping

Laminated Object Manufacturing™ uses an adhesive coated paper-type material from a roll that is pulled over a table. A laser beam controlled from CAD data strikes the material and cuts out the pattern. A heated, laminated roller travels over the material and cements the material to the previous one. The platform is lowered as repeated layers of material are added and laser cut.

After the process is completed, the laminated stack is removed, and the material surrounding the desired object is separated. The remaining pattern is of composition similar to wood, and it can be readily machined, sanded, polished, or painted. See Figures 25:7 and 25:8.

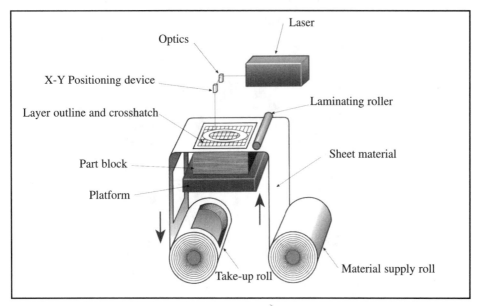

Courtesy Helisys

Figure 25:7
Laminated Object Manufacturing

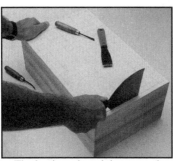

The laminated stack is removed from the machine's elevator plate.

The surrounding wall is lifted off the object to expose cubes of excess material.

Cubes are easily separated from the object's surface.

The object's surface can then be sanded, polished or painted as desired.

Courtesy Helisys

Figure 25:8
Separating the Object From the Laminated Stack

Rapid prototyping is used for silicon rubber molding to make urethane or epoxy cast plastic parts. See Figure 25:9.

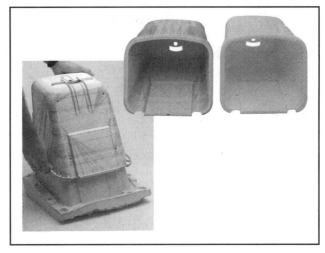

Courtesy Helisys

Figure 25:9
Silicon Rubber Molding

The rapid prototyping process can be used to manufacture solid core or core boxes for sand casting. Large and bulky patterns can be manufactured with this method as shown in Figure 25:10. The bottom picture shows a large diesel housing that was made to deliver aluminum and sand cast prototypes.

Courtesy Helisys

Figure 25:10
Rapid Prototyping and Sand Casting
Aluminum and sand cast prototype made for a large diesel housing.

Selective Laser Sintering

Selective Laser Sintering (SLS) uses a thin layer of heat-fusable powder, such as polycarbonate, nylon, investment casting wax, and metals, that has been evenly deposited by means of a roller. A CO_2 laser, controlled by a CAD program, heats the powder to just below the melting point and fuses it only along the programmed path.

The part is lowered and another layer of powder is put over the surface. Again the laser heats the powder to the point of sintering it to the previous layer. This process is repeated until the desired shape is produced. See Figure 25:11.

A manufacturer of selective laser sintering claims it can hold tolerances in the range of +/- .002″ (.051 mm) to +/- .010″ (.25 mm) across the part-building envelope.

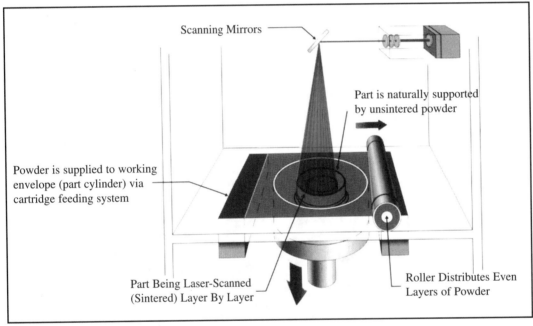

Courtesy DTM Corp.

Figure 25:11
Rapid Prototyping Using a Sintering Process

When the object is removed from the chamber, the loose powder falls away. Some post-processing, such as sanding, may be required. Patterns produced in wax can be cast into a metal of choice by using the standard investment casting process.

Sintering prototyping also allows tooling to be created from metal powder. The metal powder is a steel/copper matrix material with a thermoplastic binder. The tool produced, known as a "green part," is put into a furnace where the binder is burnt out and the metal particles are bonded. This leaves a porous but durable structure known as a "brown" part. The brown part is then put into a furnace and a second metal is used to infiltrate the porous tool. Through capillary action, the metal infiltrates the part. This produces a fully-dense tool.

Molds made with this metal exceed the physical properties of 7075 aluminum. Using proper procedures, these molds are capable of producing 50,000 or more parts. See Figure 25:12.

Courtesy DTM

Figure 25:12
Mold Cavity and Core Inserts Produced Using Metal Powder

When Woodward Governor Company was developing a next-generation aircraft fuel control system for a gas turbine engine, their engineers decided that they needed to demonstrate the new control technology. Bill Bellows, technology developer for Woodward Governor Company, said, "Our engineers decided they needed a production configuration to demonstrate the new fuel control technology, so we had to make a casting of this incredibly complicated part. We knew we had to do things differently; conventional casting was too expensive and would take twice as long as what we needed. The question was: how can we procure the casting?"

The Woodward Governor Company contacted a pattern shop that specializes in aerospace applications and complex casting. They would produce the outside core that would consist of more than 40 pieces, and the inside core would be comprised of 60 pieces.

DTM was contacted, and it was able to make the core and the mold. One of the key benefits of using their process was that no supports were used to make the core and the mold. There were substantial savings. Bellows said, "With conventional tooling, it would typically take 35 weeks just to generate the tools and another 12 weeks to get the first casting. We cut the time in half; it only took two months to get the sand cores. And the cost was 1/5 the cost of conventional tooling." See Figure 25: 13.

 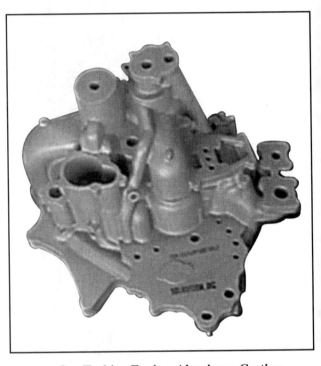

Gas Turbine Engine Core　　　　　　**Gas Turbine Engine Aluminum Casting**

Figure 25:13
A complicated fuel control system manufactured with rapid prototyping.

The 3D Keltool™ Process

The 3D Keltool™ process uses a similar process to make molds as Selective Laser Sintering. First, the desired shape is made with rapid prototyping. This master, which includes draft and .8% shrinkage rate, is then sanded and polished.

After the master is completed, it is enclosed with powered steel, usually of A6 tool steel and a binder material, and then allowed to cure. After curing, the part has sufficient strength to be handled. The "green part" is demolded and then put into a furnace to fuse the metal particles and eliminate the binder. This fusing process produces a mold composed of 70% steel and a 30% void. The void is then infiltrated with copper, producing a solid mold. Over 500,000 glass-filled plastic parts have been produced with this process. With other plastics, 10 million molded parts have been produced from a single mold (See Figure 25:14).

Courtesy 3D Systems

Figure 25:14
Hard Tooling Produced From a Rapid Prototyping Master
A PCMCIA wireless data modem housing for Research in Motion, that has been injection
molded with 50% glass-filled nylon by Durden Enterprises.

Solid Ground Curing

Solid Ground Curing (SGC) by Cubital projects a sliced image from a computer model onto a glass plate covered with toner, called a photomask. A thin resin layer is applied on a flat workpiece. The photomask is placed above the workpiece and a UV light projects the image onto the workpiece. The projected image hardens, and the unsolidified resin is removed from the workpiece. Melted wax is spread over the workpiece and cooled. The workpiece is milled, leaving a precise thin layer of resin. The workpiece travels under a powerful UV lamp for curing. A new layer of resin is applied and a new image is projected on the glass plate covered with new toner. The entire process for one layer can be finished in 65 seconds.

After the model is completed, it is encased in a solid block of wax. The wax is then either rinsed or melted away. The manufacturer claims the X-Y resolution is better than .004″ (.1 mm) and the Z resolution (layer thickness) is from .004-.006″ (.1-.15 mm). Solid ground curing is done in a solid environment which eliminates the need for supports. It also allows for the development of fully functional models with multiple parts. See Figure 25:15.

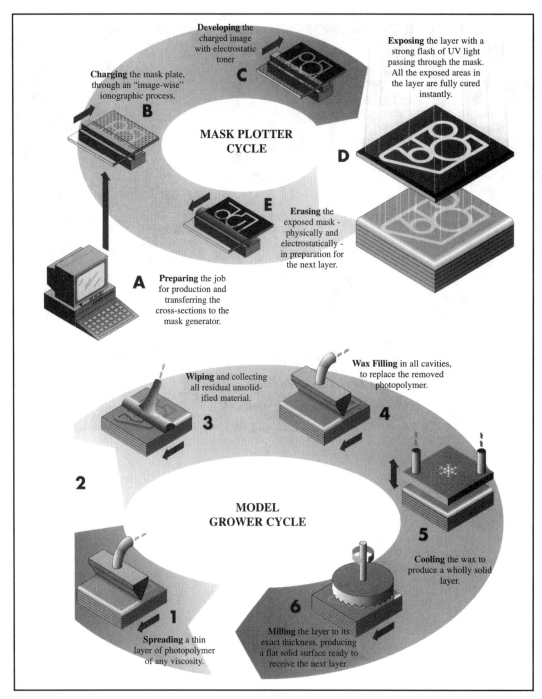

Figure 25:15
The Solid Ground Curing (SGC) Process

Fused Deposition Prototyping

Fused deposition modeling (FDM) by Stratasys is a rapid prototyping method that applies thin layers of heated, extruded, semi-liquid thermoplastic material from a CAD-generated model. FDM relies on the quick-cooling thermoplastic to solidify before applying another thin layer. Material options include ABS, medical-grade ABS, investment casting wax, and a line of elastomers.

A three-dimensional CAD model slices the object into horizontal layers. A spool of thermoplastic modeling material, .070″ (1.78 mm) diameter, is fed into a temperature-controlled extrusion head. The computer-controlled extrusion head deposits the semi-liquid material on a foam base. Each layer laminates to the previous one until the desired shape is achieved. See Figures 25:16 -17.

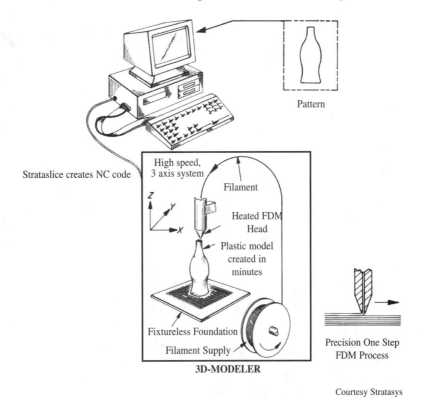

Pattern

Strataslice creates NC code

High speed, 3 axis system

Filament

Heated FDM Head

Plastic model created in minutes

Fixtureless Foundation

Filament Supply

3D-MODELER

Precision One Step FDM Process

Courtesy Stratasys

Figure 25:16
Fused Deposition Modeling

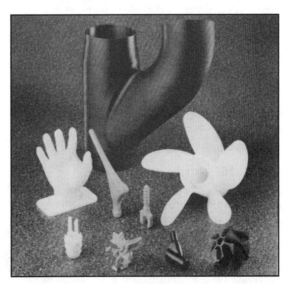

Courtesy Stratasys

Figure 25:17
Various Shapes Produced With Fused Deposition Modeling

Ballistic Particle Manufacturing

Ballistic particle manufacturing uses a modeling technique that ejects microscopic particles of molten thermoplastic that freeze upon impact. The 3D thermoplastic printer for office use uses a multiaxis head to create models from CAD designs. See Figure 25:18.

Courtesy BPM Technology

Figure 25:18
Ballistic Particle Manufacturing

The jetting system shoots the particles at speeds as high as 12,000 microparticles per second. When the particles impact the model that is being formed, they flatten and slightly melt the surrounding plastic. This creates a glue-like bond between the surfaces.

Multi-Jet Modeling Prototyping

Multi-Jet Modeling (MJM) uses a technique similar to inkjet printing. The print head of the MJM unit contains 96 jets in a linear array. As the head moves over a platform, each individual jet is programmed to apply a specially developed thermopolymer material. As the head moves rapidly back and forth, the various jets deposit a controlled amount of thermopolymer. The thermopolymer hardens within seconds. Successive layers are deposited until the desired prototype is finished. See Figures 25:19 and 25:20.

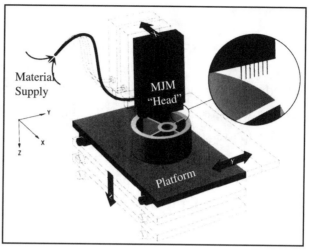

Courtesy 3D Systems

Figure 25:19
The Multi-Jet Modeling (MJM) Head

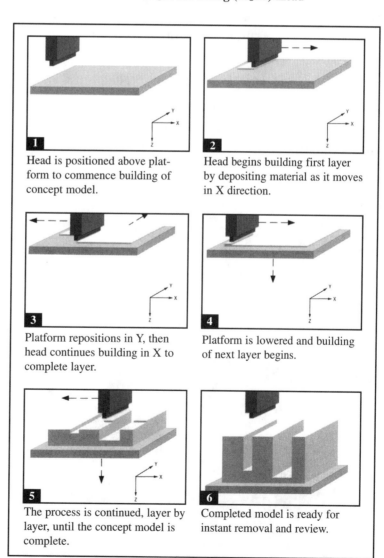

1. Head is positioned above platform to commence building of concept model.

2. Head begins building first layer by depositing material as it moves in X direction.

3. Platform repositions in Y, then head continues building in X to complete layer.

4. Platform is lowered and building of next layer begins.

5. The process is continued, layer by layer, until the concept model is complete.

6. Completed model is ready for instant removal and review.

Courtesy 3D Systems

Figure 25:20
Multi-Jet Modeling Process

The ProMetal Process

The ProMetal process uses Three Dimensional Printing, and Solid Freeform Fabrication technology invented at MIT. The process takes the 3D data and a piston deposits powered metal components layer by layer. A smooth layer of a few thousandths of an inch is deposited, and then the 2D image is printed by an ink jet printhead which deposits up to 320,000 of droplets of binder per second. After the binder is deposited, the layer is quickly dried, and another layer is deposited. This process is repeated until the 3D image is completed.

Once it is completed, the "green form" has sufficient strength to be removed, and the excess powder can be easily brushed off. This leaves a porous steel object bound with polymeric binder. The printed part is placed in a furnace to be sintered, and this leaves a porous form approximately 60 percent dense. The part is reheated and molten bronze infiltrates the porous form.

The part, composed of 60 percent steel and 40 percent bronze, is now ready for final finishing by surface texturing, additional machining, hand finishing or automated polishing.

With this process (See Figure 25:21-22), a direct metal part is produced instead of a plastic prototype. Parts with internal shapes, normally reserved for investment casting are now feasible with this process. Cooling lines in molds can be designed, allowing shorter cycle times.

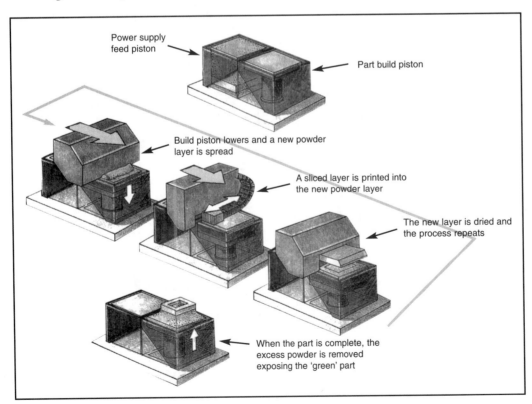

Courtesy Extrudehone

Figure 25:21
The ProMetal Process

ProMetal Rapid Tooling System

Mold in the Green Form

Finished Mold

Finished Parts

Courtesy Extrudehone

Figure 25:22
The ProMetal Process

Advantages of Office Modeling Systems

Although some modeling systems do not produce parts as strong and accurate as those built from photopolymer resins, the process is easier and faster. The great advantages of having a modeling machine in the office is that engineers can immediately produce a solid model and correct any flaws they detect in the early stages of the design.

Expanding Field of Manufacturers and Methods

Due to the ability of computers to slice three-dimensional shapes, various systems are now on the market using lasers, UV light, and plastic injectors. Rapid prototyping is expanding as manufacturers understand its great potential to reduce costs. The advantages of this process will be discussed in the following chapter.

26 Understanding the Rapid Prototyping and Manufacturing Process

Rapid prototyping and manufacturing is changing the way industry is introducing many of its new product designs. Because three-dimensional models and parts can now be duplicated from computer-drawn designs, the process offers many advantages.

Advantages of Rapid Prototyping and Manufacturing

A. Eliminates Time-Consuming and Costly Model Making

Thousands of models, prototypes, and patterns are constantly being made in many industries. Previously, these parts were time-consuming and costly to produce. Now with rapid prototyping, they can be made unattended and directly from a computer design.

When Ford Motor Company needed production units for its Explorer rear wiper motor units, rapid prototyping was used. When the first model was made, they discovered a clearance problem. They manually corrected the model and then corrected the CAD model.

New patterns were made; then investment castings were produced using A2 steel. The cost of the rapid prototyping tooling was $5,000 per tool set, compared to $33,000 that was quoted for a single tool. In addition, Ford started durability and water tests eighteen months ahead of schedule. See Figure 26:1.

Courtesy 3D Systems

Figure 26:1
Rear Wiper Motor Assemblies Made With Rapid Prototyping

In the early years of rapid prototyping, it was used primarily for reviewing prototypes of engineering design. Today the process is being used to test prototypes functionally. Chrysler used a rapid prototyping model of an intake manifold for the Dodge Viper engine, and tested it for hours at varying speeds, including full throttle. General Motors used functional models from rapid prototyping to test its heating, ventilation, and air conditioning systems.

Rapid prototyping has reduced substantial labor and tool costs. An engineering manager for a world leader in semiconductor wafer handling said about rapid prototyping, "On medium to high complexity designs we're saving from 85% to 90% in labor cost and time to develop prototypes."

A small contract manufacturer used rapid prototyping to develop a new chest clip for a child's car seat. They needed to test several designs, and multiple tooling would have cost $20,000 for each revision. With the use of rapid prototyping, they were able to test and perfect the design before going to tooling, thereby saving thousands of dollars in tooling costs.

B. Fast Turnarounds

Instead of taking weeks and even months to produce parts, rapid prototyping allows manufacturers to get parts quickly to their customers. This competitive advantage is important in today's marketplace.

A fully functional chair was produced for a manufacturer of auditorium seating with rapid prototyping. The seat was designed in four sections, two for the seat and two for the back, with reinforced rib sections. The sections were epoxied together. They were able to cut eight weeks from their development time and save over $50,000 by revealing a flaw in the initial design. By using rapid prototyping, they were able change the design quickly and save the cost of retooling. See Figure 26:2.

Courtesy Stratasys

Figure 26:2
A Fully Functional Prototype Produced With Rapid Prototyping

C. Redesigned Models Easily Produced

Designs can be tested for performance, matching hole patterns, and clearances before costly units are produced. A physical prototype can often reveal design flaws that otherwise would not be detected just from looking at a computer screen. The design can be changed easily on the computer and a new part remade and tested. See Figure 26:3.

Courtesy 3D Systems

Figure 26:3
Transparent Water Jacket Made With Rapid Prototyping
A high-performance V6 racing engine water jacket was tested using 60 sensors.

A vice president of a firm producing hedge-trimmers said this about prototyping, "The tool for our new hedge trimmer cost in excess of $200,000, so we had to be sure of the design before we committed to that kind of money for tooling. Had we used the process on our previous project, we would've found a mistake prior to tooling and really saved some money." See Figure 26:4.

Courtesy DTM

Figure 26:4
Rapid Prototyping Used to Reduce Costs for Hedgetrimmer

D. Enhanced Visualization for Design Verification

When designers and engineers can physically handle and examine an actual three-dimensional part rather than look at the same model on a two-dimensional computer screen, they have many advantages. Rapid prototyping promotes more constructive suggestions among those analyzing and distributing a full-size model at a review meeting. Such models allow engineering teams to verify critical and high-risk parts early in the design stage and to make needed improvements. One company produced three models from the original that were reduced 8%, 10%, and 12%. The design team then chose the model it thought best and revised its CAD drawings.

E. Rapid Prototyping and Investment Casting

The molded shapes produced with rapid prototyping can be used as a wax substitute for investment castings. After the molded part is constructed with gating and a central sprue, the part is put into a casting mold and investment slurry is poured around it. After the investment slurry solidifies, the mold is heated and the pattern is burnt out of the mold. The appropriate material is then poured into the mold to produce the desired part. See Figure 26:5.

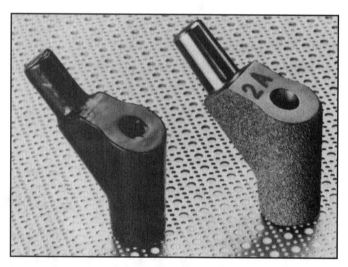

Courtesy Stratasys

Figure 26:5
Hip Replacement Joint Produced With Rapid Prototyping
(Wax pattern is on the left, casting is on the right.)

Using rapid prototyping, an auto manufacturer was able to produce a special exhaust manifold for a V-10 engine. The manufacturer saved more than 18 weeks in production time and $50,000 in tooling costs. In one year, a company made more than 4,000 precision shell investment castings without tooling, in aluminum, stainless steel, tool steel, copper alloys, inconel, magnesium, and titanium.

F. Producing Hard Tooling With Rapid Prototyping

By generating a negative CAD drawing, a pattern can be cast that will enable a fixture to be created for hard tooling for both prototypes and production tooling. Also, core and cavity inserts can be created for use in steel fixtures.

Rapid Milling for Prototypes

Rapid milling for rapid prototyping differs from CNC milling in that it relies on special high-speed milling machines to create the desired shape. For some applications it provides for better finishes, less handwork, higher accuracies, and greater choice of materials. It is said that if a machine cannot mill at 200 ipm (5,000mmpm) feed rates, it is probably not useful for prototyping.

Some of the problems associated with rapid milling are the accelerating and deaccelerating of the machine when it is traveling so fast. Computer programs need to automatically adjust the cutter path and to provide optimum feed rates with accuracy.

New Developments

In just a few years, many improvements have been made in virtual prototyping and manufacturing. One manufacturer claims that is now possible to build .015″ (.38 mm) thick walls 4.000″ (102 mm) long that are accurate within .001″ (.025 mm). Since many changes are taking place in this field, each system needs to be tested and evaluated to determine if the process, accuracies, and costs meet a company's specific requirements.

Conclusion

Virtual prototyping and manufacturing have great potential to aid manufacturers in lowering costs, and in quickly producing molded parts and tooling without the need for skilled craft people machining the desired shapes. With this technology, design engineers can produce solid 3D objects and functionally test the prototype parts from their CAD drawings.

Rapid prototyping has greatly advanced in its accuracy and in its ability to use various materials for making patterns. The future of rapid prototyping is promising in further minimizing cycle times, increasing productivity, and reducing costs.

Unit 12

Purchasing Equipment

Notes

27 Purchasing Equipment

One of the most difficult decisions for manufacturers is to determine which machining method and which brand of machine will produce the best results for their purposes. Further complicating the matter, the best results for a particular operation may not be the best process for overall manufacturing.

Purchasing a machine costing from $100,000 to $500,000 or more is an important business decision. Additionally, some machines require a significant investment in support equipment as well as training personnel to learn and operate these expensive units.

Staying Abreast With New Technology

To stay competitive, manufacturers must stay abreast of the new technology that is being developed. In 1981, someone proved mathematically that wire EDM could not achieve speeds over 4 sq. in. (43 mm/min.) per hour.

Those who experienced wire EDM in the early '80s may have decided that this process was inefficient and costly. While true then, some of today's machines are rated to cut seven times faster! For many applications, wire EDM is an extremely cost-effective machining operation.

The next important question is: Should a company purchase and train the necessary personnel to operate the equipment or should it use outside sources? This chapter deals with the decision to purchase the equipment. The next chapter, "Outsourcing" will deal with the benefits of outsourcing.

Benefits of Purchasing Equipment

A big advantage to having equipment on the premises is to satisfy customers increasing demand for "just in time" (JIT) machining. Practically all non-traditional machining methods rely on Computer Aided Manufacturing (CAM), which often eliminates the costly and time-consuming process of making hard tooling to manufacture needed parts.

A customer may need 5,000 parts in one week. To stamp the parts may be more cost effective; however, waiting for hard tooling to be made would prevent the customer from getting his order. By using laser or precision plasma for cutting, the parts can be produced within the customer's time framework.

Many customers have their own Computer Aided Design (CAD) system. They can send the CAD file over a modem directly to the manufacturer's computer. With slight modifications, the program can be loaded to a laser, plasma, abrasive water jet, or wire EDM machine. The speed and versatility of today's CAD systems helps manufacturers to satisfy customers' "just in time" delivery schedules.

American Machinist told of a firm that purchased a laser to cut various stainless steel blanks for kitchens, instead of another automatic punch press. Within six months, their laser did 50 percent of their former shearing and punching operations.

Purchasing the laser allowed them to have greater flexibility to produce special orders. They also reduced labor cost for material handling by 50%, raw-material inventory by 45%, and work-in-process by 35%.

Making the Right Selection

In today's fiercely competitive market, making the right decision is critical when choosing a machining process and machine brand. A wrong choice can have a lasting negative impact on company profitability.

A. Selecting the Right Process

Many questions need to be asked when selecting the right process. Take cutting holes as an example.

1. Does the process call for cutting one hole, a few, or many holes?
2. Is the process efficient for the quantity required?
3. Can the process cut the material?
4. How thick can the machine cut?
5. What are the required accuracies?
6. Should a traditional or non-traditional process be used?
7. Should a combination traditional and non-traditional system be used: such as a turret press with a laser or high-definition plasma?

These non-traditional processes all cut holes:

- lasers: $CO_{(2)}$, YAG, and excimer
- plasma and precision plasma
- abrasive water jet
- ram EDM
- wire EDM
- fast hole EDM drilling
- photo chemical machining.

The same goes for cutting flat blanks. These non-traditional processes all cut blanks:

- $CO_{(2)}$ and YAG lasers
- plasma and precision plasma
- abrasive water jet
- wire EDM
- photo chemical machining

A choice must be made as to which process works best. Generally, wire EDM is not for production cutting of blanks. But if 750 parts made out of .004″ (.1 mm) stainless shim stock were needed, wire EDM would be the choice. A 3.000″ (762 mm) stack could be made and the 750 parts would be cut accurately and with burr-free edges. If many holes were needed, then wire EDM would not be the choice, since each hole requires a starter hole.

Before choosing a process, make sure the process can do the work. When we owned a laser company, the person we bought the laser from began working as a salesman for a company with a precision plasma cutting system. I met the salesman at a tool show while he was handing out advertising which promoted his company's precision plasma cutting system. He claimed that his machine could do work cheaper than laser cutting.

Shortly after this, a customer of this same salesman came to our company to have their part laser cut because that company could not cut the parts successfully with their precision plasma cutting system. Not much later, that salesman began working elsewhere. The machine may not have been at fault; nevertheless, the precision plasma did not perform to the customer's expectations.

A job shop laser owner was interested in purchasing a precision plasma cutting machine. He was told that the machine could hold accuracies to +/- .005″ (.13 mm). He wisely had them cut parts. He discovered that the company's machine could only hold +/- .015″ (.38 mm).

The cheaper precision plasma cutting machines are making inroads into the sales of the more expensive laser cutting units. Undoubtedly, with increased developments, the performance gap will become closer. Lasers have their advantages, and so do precision plasmas. If there is a choice between the two systems, both should be thoroughly tested to determine if they meet the buyer's criteria before these units are purchased.

As has been shown, various processes can be used. However, vast difference of efficiencies exists between the various machining methods. If difficulties are encountered in determining which is the most profitable machining method, a buyer should go to the various manufacturers of these machines and have them do test cuts.

B. Selecting the Right Machine

A number of variables greatly affect the capabilities of non-traditional cutting machines. One important variable is the accuracy of the motion system. Although the cutting unit may be accurate, if the motion system is inaccurate, then the cut will duplicate the inaccuracies of the motion system.

In selecting the right machine, one should be aware that there can be a substantial difference between motion accuracy of the machine and machining accuracy. Manufacturers often quote motion accuracy of the machine; however, this does not mean that the machine will achieve this accuracy when cutting. For example, in plasma cutting, the motion system may be accurate, but nozzle wear, thickness of material, speed, and other factors all affect the accuracy of the cut.

Another variable for accuracy is heat generation. When heat is produced in the cutting process, it becomes an important consideration for close tolerance work due to heat expansion. The larger the part, the more critical is heat expansion.

On some cutting processes, repeated experiments are necessary to achieve the ideal settings for a particular material. Considerable time can be spent attempting to achieve the "ideal" setting. CO_2 laser cutting involves many factors that can affect the accuracy of the cut: gas pressure, focal point of lens, machine speed, and beam and lens condition.

Part accuracy is affected by the height of the material being cut. Some machines produce a substantial difference in cut quality on thick pieces. Although wire EDM is relatively slow compared to other cutting processes, it is the most accurate. With skim cuts, some wire EDM machines can cut within +/- .0001" (.0025 mm) to 4" (102 mm) and higher.

Wire EDM can do repeated skim cuts, whereas all other cutting processes rely on finishing the part in one cut. By having repeated skim cuts, extreme accuracies can be obtained with wire EDM.

Length of cut is a factor in obtaining accuracy. There can be a big difference in achieving accuracies between cutting a 1" (25.4 mm) square or a rectangular shape of 14" (355 mm) square. The longer the cut, the greater the effect are the inherent stresses within the material that is being cut. The machine may cut perfectly, but when the cut is made, the stresses within the material are relieved and the part moves. This holds true for all cutting processes.

For many cutting operations, the surface finish varies with the height of material. With some machines there is a vast difference in the surface finish on thicker cuts. Not so with wire EDM. By constantly feeding new wire into the cut, and having the new wire travel through the entire cut surface, wire EDM creates a fine finish, even on tall pieces.

Water jet, precision plasma, and laser use a cutting medium that comes out of a nozzle or a lens and tends to diffuse as it cuts through the material. Therefore, the thicker the material, the lesser the accuracy. However, these processes can be accurate with thin materials.

Listed are some of the capabilities of the various non-traditional cutting processes. These parameters will provide some guidelines to help manufacturers choose the best system to serve their needs. One word of caution concerning the accuracies of these machines. Laser, plasma, and abrasive water jet cutting accuracies can vary substantially depending on operator skills, speed, and thicknesses.

Various Machines and Their Cutting Capabilities*

A. Wire EDM

Maximum material thickness	up to 18″ (406 mm) and taller on modified machines (At our company we wire EDMed a 30 in. (762 mm) titanium bar.)
Surface finish	.35 to .45 Ram and finer
Accuracy with skim cuts	+/- .0001″ (.0025mm)
EDMing with one cut	2.00″ (50 mm) +/-.002″ (.05mm)
Material	electrically conductive

B. Laser (1500 watt CO$_2$)

Maximum material thickness	up to 1/2″ (12.7 mm) in steel stainless steel; up to 3/16″ (.4.8 mm)
Surface finish	varies with thickness: 1/4″ (6.35 mm) 250 rms 1/16″ (1.57 mm) 60 rms 1/32″ (.812 mm) 30 rms
Accuracy	+/-.002″ to +/-.004″ (.05 mm to .1 mm)
Material	electrical and non-electrical conductive, but highly reflective materials like gold, silver and highly thermal conductive materials like aluminum and copper are difficult to cut

*These cutting speeds and surface finishes are general. New technologies are constantly being developed.

C. Precision plasma (70 amp)

Maximum material thickness	up to 3/8″ (9.52 mm) steel, and 90 amp up to 5/8″ (15.8 mm) steel
Angularity of cut	up to +/- 2 degrees
Accuracy	+/-.010″ (.25 mm)
Accuracy production	+/-.015″ (.38 mm)
Material	electrical conductive

D. Abrasive water jet

Maximum material thickness	up to 8″ (203 mm) steel
Surface finish	125 rms in 1.0 to 1 1/2″ (25 to 38 mm) steel
Accuracy	small parts to +/- .0025″ (.063 mm)
Production accuracy	+/-.005 to +/-.010″ (.127 to .254 mm)
Material	From steel to granite

Choosing the Brand of Machine

After many round table discussions and testing, the process has been finally chosen. Now the decision must be made which brand of machine should be bought.

This is a critical decision. Once the final decision is made, then remaining purchases will likely be of the same brand of machine. If an initial faulty decision was made and another brand of machine is purchased, then training personnel on a different machine and learning other programming procedures can be costly.

At our company, Reliable EDM, a number of years ago we were losing jobs because of the taper limitations of our machines. We bought a wire EDM that could machine parts at 15 3/4″ (400 mm) with up to 30 degree taper. Having that machine was like having to learn wire EDM all over again due to the difficulties in learning how to program and operate the new machine. Training our operators on this new machine was costly; however, we do not regret our decision, for we bought a second machine. These machines have enabled us to do many jobs that we previously were unable to do; in fact, we can now cut up to 45 degrees taper, 16 inches (406 mm) tall. Nevertheless, before choosing a brand of machine, a buyers should do their homework carefully concerning service, capability, and reliability of the company. The consequences of choosing a wrong machine can be very costly.

A. Service

What is the company reputation concerning service? Do they readily reply and give quick attention to phone calls when problems arise? Do they have a readily available inventory of spare parts?

Often non-traditional machines are quite sophisticated. Purchasing a certain manufacturer's machine leaves the purchaser at their mercy, for only they carry many of the required parts. In addition, they are usually the only ones that can answer problems associated with their machines.

B. Capability

Does the machine perform as advertised? It is wise to check out a new machine before investing large sums of money. The buyer should always bear in mind that advertisers always want to communicate the optimum machine capabilities.

Certain machines also have optimum cutting heights. Wire EDM machines are usually rated to cut from 23-28 square inches per hour (247-301 mm/min.). This speed rate is for cutting an optimum piece of steel; it is usually a hardened piece of D2 that is about 2 1/4" (57 mm) thick. Cutting a piece much thinner or taller can greatly reduce the cutting speed.

A number of years ago I had an experience that taught me an extremely valuable lesson. I took a trip to Chicago to visit two wire EDM companies to examine their new machines. We currently had both brands in our shop. Both companies rated their cutting speeds the same. While I watched, I had each company do a test cut with the same material on their new machine. One machine cut 21 sq. in. an hour, and the other one cut 8 sq. in. an hour. Over 2 1/2 times slower! I was amazed. The person operating the slower cutting machine tried to make it cut faster, but the wire kept breaking. I expected him to give me some explanation of why it wasn't cutting better. But he never said a word. When I left that place, I was extremely thankful that I took the time to verify the manufacturers' claims. Today, most of our machines consist of the brand that cut well. That other company is now out of business.

C. Reliability of Company

What is the financial health of the company? Will they be in business ten years from now in order to service their machine? Companies do fail and go out of business. One does not want to purchase an expensive machine from a weak company, even if it has a cheaper price. For example, many of the machine's printed circuit boards electronic components cannot be found elsewhere if the machine fails. Replacing these parts can be extremely difficult if the company goes out of business.

Retrofitting or Building a Machine

Should an entire system be bought, or should a machine be retrofitted or built to a company's specific needs? I visited a profitable laser shop that had a number of lasers mounted on top of CNC milling machines. One of their particular operations involved sheared material being held in a vise as it was laser cut.

Lasers, water jets, and plasma cutting units are generally separate units mounted on a motion system. The accuracy of these machines is largely determined by the accuracy of the motion system. These motion systems can be built, or a motion system can be retrofitted to accept laser, water jet, or plasma.

Factors in Purchasing Equipment

These are some of the factors that should be considered for purchasing equipment:

A. Can the machine process the material?

B. Does the process meet accuracy requirements?

C. Is the surface finish satisfactory?

D. Does the process affect the material, for example: unsatisfactory heat affected zone, surface cracks, or burrs?

E. Is this the most cost effective method?

F. In the long term, is this the best process for various operations?

G. Which manufacturer has the best and most reliable equipment?

H. What is the financial health of the company?

I. Does the machine manufacturer have excellent service and sufficient spare parts inventory in case of machine breakdowns?

J. What are the initial as well as the additional costs associated with having this equipment in house?

K. Does the company really need to invest in this machine, or would it be wiser to invest the money to enhance other machining operations? (This issue will be discussed in the next chapter.)

Machine Installation

The decision has been made to purchase a machine. Now the proper environment for the equipment should be considered. If close tolerances are to be held, then a year-round climate-controlled environment needs to be prepared.

Sufficient space must be provided for the handling of the equipment that goes into the machines. Some machines, such as lasers and plasma cutting machines, need large areas to handle metal sheets.

To avoid unnecessary repairs, an important issue to consider are the surges and spikes of electrical equipment. These surges and spikes can come from the electrical company or from internal equipment, as in the starting of machines or welding equipment.

We were constantly repairing our wire EDM machines due to electrical surges and spikes. About every two weeks we would take the oscilloscope and check and repair the burnt out power modules. Often they were damaged beyond repair. After reading how other companies solved their equipment breakdown by installing transient voltage protectors, I installed them on my machines. There was an immediate dramatic decrease in repairs. To better protect our wire EDM machines when we had my new building erected, we had a dedicated power source installed for our wire EDM machines.

Decision Making

Machining innovations in the past few decades, particularly in non-traditional machining, have revolutionized the manufacturing process. To purchase expensive equipment or use contract shops are two of the major decision-making concerns for manufacturers. In the next chapter the benefits of using outside sources will be explored.

Unit 13

Outsourcing

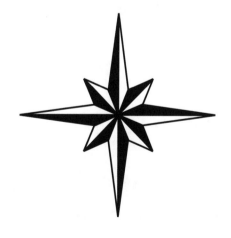

Notes

28

Outsourcing

The headline of an article in *Forbes* magazine read, "To make themselves meaner and leaner, U.S. companies keep jobbing out chores to specialists. This is one trend that will only grow in importance."

The beginning of the article states, "In the campaign to become more efficient, U.S. companies are farming out an ever greater variety of tasks that were once done in-house. Stephen McClellan, a business services and supplies analyst at Merrill Lynch in San Francisco, estimates outsourcing contracts increased by about 40% last year, up from a 23% increase the year before. 'This trend had altered the way people do business in this country,' says McClellan. 'Companies are pouring money into this industry.'"

Why are so many companies outsourcing their work instead of doing the work themselves? As a tool and diemaker who has worked in both manufacturing plants and job shops, I can testify that there is a vast difference in the mentality of workers in manufacturing plants versus job shops. Often in the manufacturing machine shop, there is no competition. With such workers, the important issue is often not how long it takes to get the job done, but whether there are few or no problems. I have seen tool and die makers wasting time standing by their machines and whittling away material. Manufacturers often tolerate such workers because they lack the experience to supervise them, and they need the tool and die makers to keep their dies and machines running.

On the other hand, job shops must quote their work. These shops must supervise their workers so that they work efficiently; otherwise, it will not be long that they will be out of business.

My son visited a large tool and die shop seeking to purchase one of their wire EDM machines. This particular shop had 60 employees, and they had four people operating four wire EDMs. The manager programmed the wire EDM. When my son arrived, he had the machines cutting at 4.5 square inches an hour (48 mm\min.). He asked the manager why the operators didn't use the recommended power settings in the machine. The manager said, "Oh no, they break the wire."

When my son inserted the recommended power settings from the manufacturer, he had the machine cutting 13 square inches an hour (140 mm\min.)—three times faster!

Advantages of Job Shops

A. Competitive Outsourcing

Job shops must be competitive to remain in business. By sending prints to various job shops for quotes, one can receive the lowest price.

B. Guaranteed Quality

Parts are guaranteed to meet print specification. Often job shops take responsibility and pay for damages.

C. Reliable Service

Many times, specialized job shops have multiple machines which can deliver quick service. This is especially important with today's stress on just-in-time machining.

D. Efficiencies of Job Shops

Job shops must remain efficient and well-informed with the latest technologies in order to remain in business. Manufacturers often have only one or two machines with one or two operators upon whom they depend. Often these operators are uninterested, or lack the expertise to keep their machines performing at maximum efficiency.

The president of of one of our competitors ridiculed our low prices for wire EDM work. Two years later, just before the Houstex SME Show, the vice president of this company (which at that time had three wire EDM machines) confided to me that they wouldn't be advertising at the show that they did wire EDM work. Instead, they were letting us do the work. This admission was a great compliment to our company that we were able to stay so competitive that they were unable to compete against us.

Costs to be Considered For In-House Work

A. Fixturing and Machine Modification

Some machining requires special fixturing in order to do the job efficiently. Our company has spent considerable time building fixtures so that we can work efficiently.

Another advantage of specialized job shops is their experience in building fixtures and modifying equipment. Our company once had a very large order for cutting titanium parts. Another company that previously provided most of the work for this oil field company had a new wire EDM, but they were unable to compete against us. They even sent parts to the manufacturer of their wire EDM, but they still couldn't compete. Our secret was modifying our machine.

B. Programming and Inspecting Equipment

Some machines have on-board programming capabilities. However, these on-board systems can be slow. Specialized CAD-CAM systems are available to speed the process.

Large job shops often have inspection equipment that enhances their operations. With wire EDM, coordinate measuring systems cannot always be used. Our shop has a video measuring system capable of magnifying parts to 325X. It is often used. See Figure 28:1.

Figure 28:1
Video Measuring System Capable of Magnifying Parts to 325X

C. Training Personnel

Generally speaking, personnel who are trained on operating non-traditional machines are highly paid, experienced workers. Sending them to school is costly and requires adequate time for them to learn the system. For many systems, it can take six months and longer to learn to operate machines efficiently. Another cost factor is the amount of scrap work that is likely to be produced from inexperienced workers.

While planning my new 18,800 square foot building, I decided to become a total precision cutting shop having wire EDMs, lasers, and abrasive water jets. I flew to the state of Washington and visited Flow International to look at their abrasive water jet machines. I then visited a number of abrasive water jet companies.

I bought a 1500 watt CO_2 laser. I thought I was extremely fortunate to hire someone with experience with the exact brand of laser that I had purchased.

My son had gone to school to learn the laser, but he was so busy doing wire EDM work, that he depended on this trained worker to run and program the machine. I wanted to get away from working in the shop in order to write literacy books for reading and math, and to write children's books that teach values. However, since lasers require much maintenance, and this laser caused so many problems, I became the laser mechanic.

In addition, the experienced worker I had hired caused us many problems. We finally decided to get out of the laser business, sell the machine, and become a strictly wire EDM specialty company. I was going to tell our laser operator about our decision to close the laser company. After taking three days off, he returned and told me that he was leaving our company and moving out of town! What a dilemma that would that have been if we had still wanted to be in the laser business. Needless to say, we were extremely delighted when we sold the laser and saw it leave our plant. Nevertheless, in the time we had our machine, I learned much about lasers .

D. Costly Support Equipment

When purchasing a machine, one should consider the costs for space and necessary support equipment. Some machines need special filtering equipment and air conditioning. Others need special handling for wastes to meet OSHA requirements for disposal.

Some machines produce fumes or dust, or both, as with lasers, plasma, abrasive water jet, and ram EDM. During milling, turning, or grinding graphite for ram EDM, dust goes everywhere, unless proper filtering systems are used.

E. Cost of Unused Machines Due to Economic Slowdown

Banks require machine payments regardless of economic conditions. When I started Reliable EDM, Houston had suffered a major economic crisis due to a flood of cheap oil from the Middle East. Oil had dropped from thirty dollars to under ten dollars a barrel. During the oil bust, successful companies went bankrupt when the economy suddenly fell because of being too deep in debt. I learned a very valuable lesson: be extremely careful about going heavily into debt.

F. Machine Maintenance

In the magazine, *Industrial Laser Review*, some representatives of various laser job shops asked questions to a panel of laser system manufacturers. One job shop representative asked, "Is there any way to improve the durability of the resonator to reduce maintenance costs or downtime? Our shop runs two systems, seven day/week, 24 hours/day, and we currently spend $4000-$5000 per machine per month for total maintenance."

Another representative complained that they have to replace their ball screws every 8-10 months. The company to whom we sold our laser had five other lasers. The day before one of the partners came to look at our machine, he had been called four times during the night because of laser problems with the night shift. Lasers can be high-maintenance machines.

We spend a considerable amount of time on periodic maintenance and repairing our wire EDM machines. When repairing mechanical equipment, the problem can be visualized; visualization is often not possible when repairing electronic equipment. We have had wires broken in the middle of a cable; situations like this are extremely difficult to solve. Another time one of our workers rewired the machine wrong. After days of downtime, we were finally able to figure out the problem.

Calling the machine manufacturer for repair services can be very expensive. Expenses often include repair person's air fare, auto, and hotel expenses. They often charge premium hourly rates for their services. Even though most repair personnel are experienced, sometimes unqualified individuals are sent. If a unit is defective, the part is sent overnight, and it requires at least an overnight stay for the individual repairing the machine.

Light coming from a laser is extremely critical for proper performance. From the manufacturer we bought a new output coupler for our laser. Knowing the output coupler controls the amount of emitted light, we checked the light pattern and discovered it was defective. When I told the company manufacturer about our situation, the repairman wanted to come to our shop to repair the problem. Instead, I bought and installed an output coupler from another company which worked just fine.

Another time we had problems with the laser machine, and a repairman came to fix it. Again, we had problems with the output coupler. Since we paid for him to come to our shop, I expected him to repair it. But he insisted he had to have an output coupler from his company. I waited until he left, then fixed the problem by simply adjusting the coupler. In addition, this factory repair person didn't even know how to clean laser mirrors—he ended up scratching the ones he had cleaned!

One hesitates to complain to the manufacturer because it is the lifeblood for your specialized machine. Without their services, one cannot exist. Fortunately, we have had very few repair personnel like this.

G. Machine and Operation Costs

Cost and profit-conscious business people should carefully calculate the costs of operating and maintaining non-traditional machines. To determine the cost, the one making the decision to buy a machine should fill in the following chart and estimate the costs for each item for five years. At the end, the figures should be divided by five to determine the yearly costs.

Five Year Estimated Costs	Cost
Cost of Non-Traditional Machine	
Loan Interest for Five Years	
Shipping	
Machine Hook Up	
Property Taxes	
Fixtures	
Inspection Equipment	
Additional Space for Equipment	
Additional Space for Support Equipment	
Support Equipment (Filter System, Fork Lift, etc.)	
Air Conditioning	
Attending Classes at Supplier's Training Center	
Weeks of In-House Training	
Months of Possible Errors From Inexperience	
Machine Maintenance	
Consumables Costs (Wire, Graphite, Chemicals, Abrasives, Gas, etc.)	
Waste Disposal (Meeting OSHA Standards)	
Breakdown of Machines	
Out of Town Repair Services	
Costs for Repair Parts	
Electricity	
Programmer/Operator Costs	
Other Expenses	
Total Costs	
Deduct Machine Value After Five Years	
Actual Cost	
Divide by 5	
Estimated Yearly Costs	

The Final Decision

By carefully analyzing all machine costs, business people can make an intelligent decision whether to purchase equipment or use contract shops. For some, purchasing the equipment is the wiser decision; for others, outsourcing is a better choice.

Before making the final decision, manufacturers should consider if it is wiser to spend the money for a non-traditional machine and its support equipment, or to purchase additional equipment with which the company is familiar. Example: If the total cost for a non-traditional machine and its accessories would cost

$250,000.00, would this money bring greater returns if it were spent expanding the present business with equipment they are familiar with?

Maximizing Core Competency

Companies need to determine the core competency of their corporation and its value to the market place. Then they should concentrate on this core competency and make the process as productive as possible. They should aim at being the best in the business in their specialty. Few companies can afford to be best in all processes, especially with the rapid changes in technology. In other words: Do what you need to do to be the best—outsource the rest.

John Mariotti, a former manufacturing CEO, president of The Enterprise Group, who works with the University of Tennessee College of Business, wrote about his experience, "I can recall vividly a benchmarking analysis I led where the result was to exit entirely from two production processes (outsource them), but to invest heavily in (or "insource") as much production as possible for two other processes. We insourced these enthusiastically because they were core processes, and we saved a bundle of money by doing so."

Increasingly, many successful companies have discovered that it is best to specialize in their core competencies and to use contract specialty shops. In the next chapter, we will discuss how to build a successful company.

Unit 14

Building a Successful Business

Notes

29 Building a Successful Business

Many dream of building a successful business. Some succeed, but many fail. What makes some companies succeed and others fail? Let us examine the principles that make companies successful.

Technological Revolution

The world is witnessing a technological revolution. In 1950, the United States had 33.7% of its work force in manufacturing jobs. Today, such workers comprise only 16% of this force. The major reason for this decline is not the loss of our manufacturing ability, but the implementation of innovative ways to reduce manual labor.

In the beginning of the 19th century, 75% of Americans were engaged in agriculture; today, less than 3% are so engaged. What transpired in our agricultural society in the reduction of its labor force is now happening in factories across our nation. This does not mean the end of manufacturing, for society will always require it. What has happened to agriculture is not that we are eating less; rather, we are much more efficient. The same holds true for manufacturing. Today's manufacturing processes are much more efficient.

Although the labor force for manufacturing has been reduced, the reduction has not been detrimental to our economy. We need to realize that productivity improvement resulting from the efficient use of manpower and equipment is the cornerstone for raising a nation's standard of living.

This trend of the reduction of the workforce in manufacturing indicates that old ways of manufacturing cannot compete in today's highly technological society. Companies must keep abreast and apply the rapidly changing technologies if they want to experience success. Even though equipment costs may be prohibitive for in-house use, experienced contract shops should be utilized to perform the needed work.

One summer I taught a junior achievement economics class in English at Prague University to Czechoslovakian high school and university students. Three years previous to my coming, these students were under communist domination.

The communist countries produced the following cars: Romania—Dacia, East Germany—Trabant and the Wartburg, USSR—Lada and Moskvic, and Czechoslovakia—Skoda. A student told me that when the Czechoslovakians wanted to produce a better car, the Russians forbade them! The Russians feared that their cars wouldn't sell in the Czechoslovakian market.

We know what the market forces eventually did to the communist empire. These same forces operate today. Those understanding and practicing successful economic principles survive; those who don't eventually fail.

At age 57, I started Reliable EDM with my two sons. In less than ten years, we built an 18,800 square foot building and became the largest wire EDM job shop west of the Mississippi River. I would like to share some of the principles that I have found successful.

Author's Background

Before starting Reliable EDM, I had worked as a tool and die maker; precision machinist; foreman of a tool and die shop; operations manager for a large tool, die, and stamping company; New York City high school teacher; and assistant dean of boys. During that time, I worked ten years researching and writing the book, *Schools in Crisis: Training for Success and Failure*?

I enrolled at New York University to take some courses required for teaching high school. At New York University I had the privilege of using Bobst Library, the largest research library in New York City. I investigated the principles for producing successful schools. My research and years of experience taught me many lessons on what to avoid and what to apply. Whether trying to raise a successful family (I have five children and seven grandchildren), enjoy a happy marriage (happily married since 1956), run a profitable business, operate a high-achieving school, or whatever other enterprise one may enter, certain basic principles produce success.

Beginnings of Reliable EDM

When I started Reliable EDM, I supplemented my savings with borrowed family money. I bought one used wire EDM machine, and for $10,000 purchased two very old Agie wire EDM machines and a Hewlett Packard Computer system. We needed the computer system to program our wire EDM machine.

We managed to get one of the old Agie wire EDMs working, but the programming system was so archaic and the machine cut so slowly that we never cut one part with the machine! Although we started with three wire EDMs, we did all our work with one. To keep the one machine running at night, I put a bed and an alarm clock in one of the office rooms. During the night, I'd wake up and keep the machine operating.

I also bought a used milling machine, lathe, and surface grinder. In case we got slow, I figured I could at least support myself by doing machine shop and tool and die work. Seven months later we became so busy that we bought our second wire EDM machine. As the needs arose, we purchased more.

Philosophy of Success

I have a firm belief that there is a God, and that the Bible is God's infallible revelation to us. Because of this belief, I decided to do a detailed Bible study concerning principles for building a successful company. Whatever truths the Bible taught, I was determined to obey them. I've developed a very simple philosophy: God is much wiser than I am. If my beliefs conflict with his instructions, I disregard my beliefs and obey God's Word. I have proven over and over again that God's solutions are always best.

In my study, I have discovered two foundational truths for building a successful business: the Golden Rule and loving people.

The Golden Rule

Jesus taught his disciples, "So in everything, do to others what you would have them do to you.[1] The Golden Rule simply says: Whatever you do, love people by putting yourself in their place, and then treat them the way you would like to be treated.

Loving People

The second important foundational truth is to love people. An expert in the Old Testament law came to Jesus and tested him with this question, "Teacher, which is the greatest commandment in the Law?"

Jesus replied, "Love the Lord your God with all your heart and with all your soul and with all your mind. This is the first and greatest commandment. And the second is like it: Love your neighbor as yourself. All the Law and the Prophets hang on these two commandments." [2]

Jesus taught that the greatest commandment is to love God with all our heart, soul, and mind, and the second greatest commandment is to love people as ourselves. The two most important principles of life are to love God and people.

Applying the Golden Rule and Loving People

When I started my business, I studied the principle of the Golden Rule and of loving people. In reality, the Golden Rule is also about loving people; it is a willingness to identify with others.

To apply the principle of loving people, I asked myself, "If I were a customer, how would I like to be treated?" I put myself into my customers' shoes. In order to please my customers, I put down these three items I felt my customers wanted.

A. Quality Products

There is absolutely no substitute for delivering quality products to customers. A business may have excellent service and the cheapest price, but if the quality is missing, customers will never be satisfied.

Producing quality products is simply giving customers the products that meet or exceed their expectations. Producing quality products must be the highest priority for a business to become successful. A system needs to be implemented to ensure the delivery of high-quality items to every customer all the time.

When a business starts, often the owner is actively engaged in every aspect of the business. Problems arise when the company expands and others are hired who need to be trained to produce quality products. For the company to succeed, the business owner *must* be committed to delivering items to customers that meet or exceed their quality standards. This calls for training and supervision of workers.

I have discovered that the best place to insure quality is not in the inspection department, but in teaching each worker as he is producing the parts that he is accountable for producing quality work. In other words, every worker is an inspector. Make the training of workers a high priority. Give every worker clear instructions on how to perform their jobs, and what needs to be checked to insure the production of quality products. Avoiding mistakes is much wiser than catching mistakes.

B. Reliable Service

What is service? Service is delivering the product when the customer wants it. Customers don't like to be told, "We can deliver your product in two to three weeks," when they know it takes only two days to do the job. Apply the Golden Rule: How would you wish to be treated if you were the customer?

Houston has an exceptionally successful furniture store that sells quality furniture in a most unusual place. This place of business was once a show lot for model houses. The new owner connected the houses and filled the rooms with furniture. Though the rooms are air-conditioned, the whole front of the store is without air conditioning, almost unthinkable in the hot and humid Houston summer.

The owner advertises extensively on television that his store will, "Save you money." Anyone purchasing a store item is guaranteed same day delivery. One cannot get better service than that. The furniture store has succeeded spectacularly, expanded their operation, and built a large air-conditioned building on the premise.

In addition to delivering quality products, delivering service must be made a high priority. Today, providing service has become extremely important because many companies are practicing "just in time" (JIT) production. These companies have discovered that storing inventory is expensive. So instead of having inventory, these companies are looking for vendors providing quality products when they require them.

At times our customers have overloaded us with work. Even though we're committed to provide service, we have reached our capacity. At times like this we make plans to purchase additional equipment. (We have now have over 20 wire EDM machines.) Delivering service is critical to growth.

C. Value

It is commonly said, "You can have quality, service, and value; but you can pick only two." There is absolutely no substitute for quality, but there can be a toss up between service and low prices. Some customers will choose service, and some low prices. But contrary to the popular saying, when I started my company, I aimed to provide all three: quality, service, and value.

In the beginning I spent considerable time designing and building fixtures to increase our productivity. To this day I still build fixtures to maximize the efficiency of our machines. Why do I do this? Because this allows our company to do things faster so we can stay competitive. Imagine a company that aimed to deliver quality products, provide reliable service, and had competitive prices— wouldn't you want to do business with that company?

When I did my intensive Bible study on how to build a successful business, this is what I wrote, "If we could use one word for quality, service, and competitive, it would be: Love your customers. Let them know in no uncertain terms that you are there to help them. Customers will beat a path to any business that practices real love." And customers came.

A number of years ago at the Houstex Industrial Show, our company rented a space next to *EDM Today's* booth. *EDM Today* publishes the only magazine devoted to the EDM industry in the United States. The owners of the magazine attend tooling shows throughout the United States. They had an experience at this show that they never had at any other trade show. At other trade shows, business personnel would stop at their booth and show interest in purchasing EDM equipment. However, people in this trade show told them they knew about EDM, but they subcontracted their work. Jack Sebzda, editor-in-chief of *EDM Today*, told me that our company was like an EDM center: companies would rather come to us for their EDM work than purchase the equipment.

It showed that when a company provides quality, service, and value, customers are reluctant to buy specialized equipment. However, since then some of our customers have bought their own machines. One of them bought a machine so they could bolster their image by telling a large customer of theirs that they had wire EDM capacity.

The Most Important Principle

Successful companies aim to make their customers "number one." No business will ever succeed without pleasing or loving their customers. Speaking of love may sound mushy; nevertheless, love is life's most important principle. Everyone craves it—your mate, children, friends, employees, and customers. Master the concept of loving others, and you'll discover life's richest rewards. Be selfish and unloving, and you'll suffer the consequences.

If we could use one thought for providing quality work, with reliable service, at great value, it is—love your customers. Let them know that you are there to help them to become successful.

Teach your customers how to save money. Don't exploit them. The Bible says, "Dishonest money dwindles away, but he who gathers money little by little makes it grow."[3] Some people are so bent on becoming rich that they will do anything to achieve it. They may experience temporary success, but eventually they fail. By gathering your money little by little and treating your customers well, you'll discover the road to success.

Since I had years of experience as a tool and die maker and operations manager of a large tool, die, and stamping shop, I'd show tool and die shop owners how to build high-quality dies from one piece of tool steel without the need of aligning the die. These dies would save them thousands of dollars. What was I doing? I was trying to help them become successful. I even wrote a book with the help of my son, *Wire EDM Handbook*, which describes how to build these high-quality inexpensive one-piece dies.

What are the qualities that make companies and stores successful? Imitate them. One of the practices of successful stores is: "Satisfaction Guaranteed." I delight in going to these stores, for I know that if I'm not satisfied with what I have purchased, they will readily return my money.

Time-Tested Principles for Success

Here are some of the time-tested principles for success. They were written three thousand years ago, and they are just as true today as they were then. These principles came from King Solomon—the wisest king that Israel ever had. The Bible tells us how Solomon acquired his wisdom. The Lord asked him when he was king, "Ask for whatever you want me to give you."

This was Solomon's answer, "I am only a little child and do not know how to carry out my duties. Your servant is here among the people you have chosen, a great people, too numerous to count or number. So give your servant a discerning heart to govern your people and to distinguish between right and wrong. For who is able to govern this great people of yours?"

Solomon's request so pleased the Lord, that the Lord said, "Since you have asked for this and not for long life or wealth for yourself, nor have asked for the death of your enemies but for discernment in administering justice, I will do what you have asked. I will give you a wise and discerning heart, so that there will never have been anyone like you, nor will there ever be. Moreover, I will give you what you have not asked for—both riches and honor—so that in your lifetime you will have no equal among kings. And if you walk in my ways and obey my statutes and commands as David your father did, I will give you a long life."[4]

Solomon asked for wisdom to help others. And because he asked for wisdom instead of satisfying his selfish pursuits, the Lord answered his prayer. Today we have the book of Proverbs which contain many of Solomon's words of wisdom. Carefully consider these principles for success which I have taken from King Solomon writings.

Hiring Workers

It is crucial in building a successful company that you make sure the individuals you hire have the potential to become good workers. Solomon said, "Like an archer who wounds at random is he who hires a fool or any passer-by."[5] If a warrior shoots his arrows at random, he may hit an occasional target. But how much more successful he will be if he takes careful aim with his bow and arrow! Take the necessary and sometimes painful and costly steps to insure that you obtain good workers.

In my business, finding people experienced in operating wire EDM equipment is highly unlikely. My main concern is that new workers understand basic mathematics, are careful with numbers, energetic, and teachable.

I am particularly concerned that the people I hire are teachable. I've worked many years in different shops as a tool and die maker, and I've seen quite a number of unteachable workers. I determined that in my company I will only hire teachable workers . Everyone I hire, I tell them I want to be able to go to any worker in the shop, and if I see a machine isn't operating properly, I'll show them how. If any worker takes offense at my trying to help them, I don't want them working for me. I tell them that we don't operate a high pressure shop; rather, we stress efficiency of machine operation.

I give mathematics tests to all applicants. This simple math test, which includes using a ruler, has helped me weed out many unqualified workers. With today's educational system, having a high school diploma is no guarantee that a person knows even simple math. After a worker appears qualified, we interview them to determine their work attitudes.

If the worker appears to have proper qualifications, then we give him a simple test to examine his mechanical ability. We have him go to a mill and do a simple job of taking an indicator and lining up a vise. Then we usually do two other tests with indicating on the mill. Also, by just watching how he cranks the mill handle, we can can get an idea of how energetic he is as a worker.

Gaining Knowledge

Give top priority in your life to gaining wisdom. Spend the necessary time and money to obtain knowledge. Listen to the instructions that Solomon gave concerning this issue:

> Choose my instruction instead of silver, knowledge rather than choice gold, for wisdom is more precious than rubies, and nothing you desire can compare with her.[6]
> The heart of the discerning acquires knowledge; the ears of the wise seek it out.[7]
> Buy the truth and do not sell it; get wisdom, discipline and understanding.[8]

If you plan on going into business, learn as much as possible about the business you are starting. Make it a practice to attend tool shows and examine trade magazines. Keep abreast of what's happening in your industry. If you are in business, think and make plans on how you can make your operations more productive.

Always be alert to find ways to increase productivity. Periodically search through tool company catalogs to know the tools that are applicable for your trade. I've done this many times and bought various tools to increase the efficiency of my company. Time is money.

I've been amazed how inefficiently companies can operate. At the company where I became operations manager, press operators would spend vast amounts of time searching for tools, like a hex wrench. The cost of the wrench could be covered with less than one minute of press time, but a worker might spend ten minutes or more searching for one. It was very costly to have power presses idle while workers went around the shop searching for tools. I changed the situation by buying each operator a tool box and their own tools. For special tools, I had a tool rack built.

I've seen high-paying tool and die makers with a hack saw cutting off bolts. We could probably buy a box of bolts for the time they spent. I bought a rack with drawers for all the bolts, dowel pins, and other items we needed. I would then periodically search through the drawers and order items that were low.

I've also seen the boss on numerous occasions where I was employed drive a considerable distance just to buy items that should have been stocked. His trips were extremely costly. How much wiser to spend the money and stock up on items. Within six months, the time saved would cover the cost of stocking up.

At the company where I was operations manager, we would constantly search for stamping dies. To remedy the situation, I spent considerable time numbering each die and giving each one a home. I made operation sheets indicating where the die could be found and how the parts should be made. One sheet always remained in my office, and a copy was given to the operator. I also painted two of the parts that were stamped. One part was painted blue, and the other one white. The blue part hung in my office, and the white part went with the die. Now when we received a repeat order, the operator would receive an operation sheet which showed him the location of the die, the sample part, and how to set up the press and run the part. If there was anything unusual about the setup, I would take a Poloraid picture and insert it with the operation sheet.

I cannot complain if a worker spends ten minutes looking for a tool if the company doesn't insist on an orderly shop. No one likes to hunt for tools, but if the company doesn't care, the workers will just spend time searching. They are getting paid whether they are doing productive work or searching.

Make it a high priority to have an orderly shop. Your workers will be happy, and you will increase your productivity. Make it your motto: A place for everything, and everything in its place.

Seeking Advice

Make it a practice to seek advice. This is so important that I have included six proverbs of Solomon. You may feel like saying a big "Amen" after reading them. Here are these gems:

> The way of a fool seems right to him, but a wise man listens to advice.[9]
> Plans fail for lack of counsel, but with many advisers they succeed.[10]
> Listen to advice and accept instruction, and in the end you will be wise.[11]
> The purposes of a man's heart are deep waters, but a man of understanding draws them out.[12]
> Do you see a man wise in his own eyes? There is more hope for a fool than for him.[13]
> He who trusts in himself is a fool, but he who walks in wisdom is kept safe.[14]

I'll relate some incidents on how I tried to follow Solomon's instructions. Since starting Reliable EDM, I've always gone to Chicago to IMTS (International Manufacturing Technology Show), the world's largest tooling show, to keep abreast of the latest technology. To write this book, I've gone to various booths at the IMTS and obtained information and sought their counsel. I've searched through various trade magazines. I also had numerous experts in non-traditional machining read chapters dealing with their specialty. I've found that people are often willing to help and share information.

One of my favorite verses is, "The purposes of a man's heart are deep waters, but a man of understanding draws them out."[12] You'll encounter individuals who will have much valuable information that will benefit you. How can you obtain that information? Only by extracting it from them. A wise individual masters the art of searching out information from others. In simple words, "Learn to ask questions." I like the Chinese proverb, "He who asks a question is a fool for five minutes. But he who doesn't ask a question, remains a fool forever." So take your bucket and draw from the wells of others. You will be surprised what you can learn.

For example, before I purchase a new type of equipment, I try to learn as much about it as practical. I ask the opinion of others about what they think of it. I try to get manufacturers of other models to tell me of their advantages compared to the equipment that I intend to buy. Sales personnel often say they don't like to knock their competitor, but they usually do so subtly.

Before making major company decisions, I also seek the advice of my wife and my sons. No man is an island. But I like Solomon's version much better: "He who trusts in himself is a fool"[14]

Hard Work

Talkers and dreamers all fail. Success comes only to diligent workers. Although the path of diligence isn't easy, those willing to pay its price will reap its benefits. Trying to achieve success without doing hard work is an idle dream. Solomon wisely said,

Lazy hands make a man poor, but diligent hands bring wealth.[15]
He who works his land will have abundant food, but he who chases fantasies lacks judgment.[16]
All hard work brings a profit, but mere talk leads only to poverty.[17]

Hasty Decisions

Get-rich-quick schemes are abundant. Some experts claim by reading their literature and practicing their principles you'll discover the path to financial success. However, you must first purchase their materials to gain their secrets. You'll probably discover that the road to quick riches is littered with failures.

When I was a boy, my friend challenged me, "I'll stand across the street, and I'll put you on the ground in ten seconds."

Since I was a street kid, there was no way he was going to put me on the ground in ten seconds. So I said, "I dare ya!"

I bet him a dime that he wouldn't put me down on the ground in ten seconds if I were across the street form him. Back in the early 40s, a dime represented a lot of money—we could buy two ice cream cones, and a large pretzel cost only one cent. That wise guy took a piece of chalk and wrote the letter "U" on the sidewalk! The result? I had to pay him a dime. That incident left an indelible lesson on my mind to this day, worth much more than the dime I lost.

Give sufficient thought before making decisions. Don't be gullible. Think before you act. Solomon gave this advice, "A simple man believes anything, but a prudent man gives thought to his steps."[18] He also said, "Do you see a man who speaks in haste? There is more hope for a fool than for him."[19]

Taking Correction

Some people are so proud that the only way to get them to change their ways is to try to plant the idea into their head so that THEY think they originated the idea. Being open minded and willing to take corrections signifies a great mind. In my lifetime, I have found very few such individuals. Yet, try telling someone they are close-minded or unwilling to take corrections. There will be a violent reaction.

I once sat in a meeting where the president told his workers how open he was to their ideas. It amused me, for I happened to be the operations manager; repeatedly he refused to listen to suggestions I offered. When I began working, he eagerly listened to my ideas. But as time progressed, he began rejecting my ideas without even analyzing them.

Here is a quick test to evaluate how open-minded you are. This simple test will involve your mate; however, to avoid your mate from going into shock, have him or her sit down. Take a pad and say sincerely, "Honey, tell me my faults. I'm willing to change whatever is wrong." Of course, be ready to receive and act upon what has been requested of you.

You may ask, "But how does this relate to building a successful business?" Plenty! This test will reveal how open-minded you are. If you can't listen to the corrections of your mate, you're not open-minded, even if you think you are. Besides, whatever you do, don't neglect your mate and family in aiming for success. Having a happy mate and children is much more important than having a business that makes lots of money. Solomon said:

> A mocker resents correction; he will not consult the wise.[20]
> He who listens to a life-giving rebuke will be at home among the wise. He who ignores discipline despises himself, but whoever heeds correction gains understanding.[21]

To receive input from workers, set up a suggestion box to receive worker input. Also, go around and speak to each worker and ask for their opinions on how to make the company more productive. However, for it to be successful, one needs to value worker's suggestions.

Likewise in your home. Be open to family suggestions and evaluate carefully what they say. Remember the Golden Rule? Include your entire family in your decision making.

Experience

Solomon asked, "Do you see a man skilled in his work? He will serve before Kings: he will not serve before obscure men."[22] If you start a business make sure you are equipped with the necessary skills.

Learn as much about the industry as possible so you will be intelligent and efficient in running the operation. Before we opened our wire EDM company, a worker who managed the wire EDM shop where I was operations manager bought a new wire EDM machine and started his own wire EDM company. Shortly after starting his business, he sold the machine. No doubt to him, Houston was a terrible market for a wire EDM company.

When this worker started doing the wire EDM work for our company, I told the president of the company to replace him for his unteachable attitude. At first the owner refused, but after a year of low performance, he reassigned him to the tool room and hired my son.

What went wrong? As operations manager I tried to instruct this worker, but he wouldn't listen. He also lacked the skills for using the mathematical capabilities of the computer. Instead of mastering the computer, he'd tell customers to bring drawings ten times larger than the part in order for him to digitize them.

Two and a half years later when we started our business, we discovered an excellent market for a wire EDM company. The difference? We had much experience before we started our business. I had many years experience in the machine shop and in the tool and die making trade; and my son, a mechanical engineer and a tool and die maker, had managed a wire EDM shop for two years. And most of all—we followed biblical principles, and the good Lord was with us.

Permanent Diligence

Finally you have reached the pinnacle of success—now you can relax. Try it, and you'll soon discover the road to failure. Economic forces are always eager to challenge your successful operations. This is what makes the free enterprise system so unique and powerful in that it creates the most competitive and productive environment.

Let a successful company become easy-going and inefficient, and other companies will rise and put them out of business. These other efficient companies will underprice or out-serve the former successful company. Listen to Solomon's wise counsel, "Be sure you know the condition of your flocks, give careful attention to your herds; for riches do not endure forever, and a crown is not secure for all generations."[23]

Some business owners take a permissive approach in running their business. When I was working as a tool and die maker in Long Island City, New York, I saw how this company that I worked for went from a three-man shop to a twenty-four man shop, and then out of business. The manager of the company never corrected the workers. He believed if you treated the workers "right," they would work right. The downfall began when the foreman of the shop fired a machinist, and the manager rehired him. From then on the foreman didn't care about the shop or how the workers performed. The company continued going downhill, and shortly after I left, it closed down. The problem I saw was the manager would not correct his workers. Permissiveness destroys. To run a successful company, owners must be willing to confront problems and take corrective measures. Such action can be very painful.

The president of another company where I was operations manager, brought in his son-in-law and made him stamping shop manager. My new job now was supervising the tool room. For years the tool room was unproductive; I wanted it to become profitable. I had years of experience as a tool and die maker and as a machinist. I told the workers that I wanted the tool room to be a pleasant workplace, but also to be competitive. I told them that I planned to supervise actively, by looking at and questioning their operations.

I told one of the long-term workers how he could make the mill run more efficiently. He objected that I spoke to him about machining, so he went to the president and complained. Guess what the president did? I was corrected!

Shortly after this I told the president that my son, who was the only one operating the wire EDM department, and I were leaving and starting a wire EDM shop. I gave him a five-week notice, even though he would be one of our competitors. I gave him time so my son could train one of his workers. The president then took over the quoting, and I began working in the tool room. Now I could observe the workers.

The vice president said to me, "We ran you off." Then he told me about this worker whom I had tried to help. The company had quoted 190 hours for five dies to build, but this worker took 290 hours. He asked me what happened.

I said to the vice-president that I have years of experience as a tool and die maker, and that it would have been good for the president to strongly back me up and state to the workers, "I want Carl to go around and see how we can make this shop more productive. If he sees one of you on a machine, and it can run more productively, I want him to go up to you and show you what to do. He may be wrong. But we are all here to learn. I want you also to go up to Carl and tell him if he is doing something wrong. We don't want any hurt feelings."

Then I told him how I saw this worker take a small piece of tool steel and put it on the surface grinder. He put on the automatic feed and stood there and watched it grind. The stroke of the machine was about four times longer than the part! I told him I could have ground that piece five times faster by hand.

I said to him that it puzzled me in that how that both of you (him and the president) know how sensitive this worker is, but when he complained about me, who got into trouble? It was I! While I was speaking to him, I saw that what I was saying was beginning to sink in. Then he said that he really wanted me to stay.

He told me an interesting story about his dad who started this shop. Before 1970, this company was just a tool and die shop. In contrast to their running the company, his dad made money running the tool and die shop. His dad went around looking for problems. Then I asked the vice-president, "Was your dad like me?"

"Yes," he replied.

I've experienced firsthand the failures of permissiveness. I also saw it as a teacher when I substituted in 27 different schools in all the five boroughs in New York City teaching all grades from 1 to 12. In my book, *Schools in Crisis: Training for Success or Failure?*, which took me ten years to write, I came to the conclusion that to have successful schools, leaders must be fair, firm, and loving. This works in schools, homes, churches, businesses, everywhere.

Volumes of pseudo-psychological studies try to defend permissiveness. I've spent years researching the problem of our schools. Just look at our schools today and witness the fruits of permissiveness. Permissiveness is a cancer that destroys whatever it dominates.

That's one reason why when I started my business, I determined to have a disciplined shop, hire teachable workers, and confront problems whenever they arose. Notice in the following verses Solomon's comments, and also how those whom God loves he disciplines and rebukes. You won't go wrong following God's methods.

> My son, do not despise the Lord's discipline and do not resent his rebuke, because the Lord disciplines those he loves, as a father the son he delights in.[24]
> Whoever loves discipline loves knowledge, but he who hates correction is stupid.[25]
> He who ignores discipline comes to poverty and shame, but whoever heeds correction is honored.[26]

Keeping Abreast With Technology and Striving to be Unique

Having productive workers is critical to having a successful company, but keeping abreast of what's happening in your industry and being willing to invest company profits to remain competitive, are also vital issues to stay successful.

A number of years ago, I flew to Chicago to go to the Quality Expo International, the world's largest show for quality. I wanted to examine the various coordinate measuring machines (CMM). I had listened to various salespersons who had visited our shop, but I wanted to be sure to get the CMM with the best and latest technology. The first day in Chicago, I examined the various CMMs. The second day, I had my son fly up to finalize what I thought was the best CMM. After examining many CMMs, we finally bought a computer numerically controlled CMM which had the latest and best software.

Another time we were interested in purchasing a video-measuring microscope. We had our mind set on a certain model. But to make sure we had the latest technology, I flew again to the Quality Expo International. Optical Gaging Products (OGP) had just introduced the Smartscope. I went to the other vendor and carefully compared his machine with the OGP machine. I bought the Smartscope. Later on OGP told us that they had sold over one thousand of these machines, and that we were the first ones in the world to purchase the Smartscope.

To increase our productivity, we spent over $250,000 for a large submersible wire EDM machine. It was the latest machine on the market. We had taken in a job that was slightly larger than the machine capacity, so I modified the machine so we could cut 18 inches (457 mm) in a submersible condition. We modified another machine to cut 38 inches (965 mm). No machine on the market can cut that tall.

Wire EDM is like a band saw, the wire must go through a part for it to cut, but we build a device so we can cut one cavity in a tube. We are willing to experiment and do things that are unconventional.

To be successful, one must keep abreast with technology in order to stay competitive, be willing to be unique, and inform others about your capabilities. But how are people to know about the company's capabilities?

Establish Your Credibility

Your best advertisement is your satisfied customers. Satisfied customers tell others. But how do you get customers when you're starting a business? In the beginning of our company I phoned machine shop personnel in town and told them about wire EDM. I sent them information on the capabilities and proper procedures for wire EDM. I mentioned that my son was a mechanical engineer who did tool and die work and had managed a wire EDM shop, and that I had been operations manager of one of the largest tool and die shops in Houston and had years of experience as a tool and die maker.

What was I doing? I wasn't bragging; I was trying to establish our credibility. Solomon said, "A good name is more desirable than great riches; to be esteemed is better than silver or gold.[27]" To obtain a good name, one has to earn it.

For example, I went to a heat treater to have some parts heat treated. When he told me he had worked as a tool and die maker, I had utmost confidence in his ability to heat treat the parts properly. The fact that he was a tool and die maker told me that he had certain qualities.

Our office has two enclosed displays and our foyer has one showing many different and unique jobs produced with our wire EDM machines. If we didn't have such displays, when a person with little wire EDM experience visits our company to have a part machined, he would likely ask, "Can you make this?" But now if that same person comes and looks at any one of our displays, he would probably ask, "How much does it cost?" There is a world of difference between those two questions.

Instead of having just a plain fax letter head, our letterhead states: "Largest Wire EDM Job Shop West of the Mississippi" Then I cite our capabilities. When we started our company, I tried to get an unique 1-800 telephone number for our company. We managed to get this toll free number: 1-800 Wire EDM. Also, whenever a new customer comes to us, we usually give them a free copy of our book, *Wire EDM Handbook*, written by my son who is a mechanical engineer and myself. Being authors helps establish our credibility. We try to take advantage to advertise our capabilities without being boastful.

Having Proper Priorities

To succeed in business, one needs proper priorities. What should be done with profits? Spend it on things you have often desired but could never afford? Solomon advises, "Finish your outdoor work and get your fields ready; after that, build your house."[28] In other words, since your business is your livelihood, take care of it first before you build that house or buy that new car.

Build A Company With Character

One of the fastest ways to destroy a company is to treat customers and workers dishonestly. Abraham Lincoln said, "You may fool all of the people some of the time, and some of the people all of the time, but you cannot fool all of the people all of the time."

Be dishonest with workers, and you'll lose their respect. Remember, workers as well as customers have excellent memories. Solomon wisely said:

> Better a little with righteousness than much gain with injustice.[29]
> A fortune made by a lying tongue is a fleeting vapor and a deadly snare.[30]
> The LORD abhors dishonest scales, but accurate weights are his delight.[31]

It's a shame, the loss of character we find in our culture. Today, many are interested in making money at any cost. Men and women are desperately needed who will determine to be people of character first, above making money.

Don't believe the lie that one can't be honest and be in business. You definitely can be honest in business. When I started my business, I determined I would pay all my taxes. Yes, there are a thousand ways to cheat the government—but I vowed to be a person of integrity. One day, we'll all have to give an account to the Great Judge. I want to hear these words when I stand before his judgment seat, "Well done, good and faithful servant! You have been faithful with a few things, I will put you in charge of many things. Come and share your master's happiness!"[32]

Are there people who prosper by being wicked? Unfortunately, yes. Some individuals witness this wickedness and want to follow suit. But here's Solomon's warning, "Do not fret because of evil men or be envious of the wicked, for the evil man has no future hope, and the lamp of the wicked will be snuffed out."[33]

When Trouble Strikes

Not to expect trouble is a fool's dream. Solomon advised to go to the ant and "consider its ways and be wise! It has no commander, no overseer or ruler, yet it stores its provisions in summer and gathers its food at harvest."[34] The ant knows winter is coming and wisely prepares for it.

Before I started my business, I talked to individuals who had lost their businesses during the oil crisis in Houston. When I moved to Houston in 1978, businesses were booming. Because the oil cartel had raised the price of oil, nations could now afford oil exploration. When they began discovering oil, the oil cartel anticipated an oil glut, and drastically reduced the price of oil. Oil became so cheap that buying foreign oil was cheaper than retrieving oil from some domestic wells. Suddenly, oil companies stopped purchasing equipment, drilling, and exploration.

During that time I was operations manager and needed to hire a worker. I dared not list our company's phone number in the ad. Instead, I listed a P.O. Box to avoid being inundated with phone calls from laid-off machinists.

A large machine shop owner told me that during the oil boom an oil company promised him work if he'd get more machines. He purchased more machines. When the economy slowed down, the oil company no longer purchased his parts; but he still had to make payments on the machines he had bought. For three years he and his partner tried their best to keep the company going, but eventually they failed. Unfortunately, others failed as well.

I learned a valuable lesson from this: prepare for trouble by building from strength. So when in business, be extremely careful about going too heavily into debt. One needs to protect his company when trouble strikes—for it will strike. Solomon said, "If you falter in times of trouble, how small is your strength![35]

You'll also encounter trouble with workers. First of all, I have learned that the best way to avoid trouble with workers is to be extremely careful whom you hire. Then if you do have trouble with a worker, confront the problem. Take intelligent action and be willing to make the difficult decisions to correct problems. To avoid correcting problems only worsens the situation.

If some part of the business is unprofitable or not working properly, learn to stop the venture and count it as a learning experience. Don't be so proud that you would never retreat from a bad decision. A retreat can turn out to be an advance in a new direction.

Shortly after starting our company, I wanted it to become a total EDM job shop. To achieve this goal, I bought a new ram EDM. From the very beginning of our company, I desired my two sons to manage the EDM business so I could write books and start a publishing company. Unfortunately, although I hired others, I did much of the design and spent considerable time operating the ram EDM. Ram EDM took so much of my time and was so competitive that I finally decided to get out of this type of EDM.

A few years later, we decided to be a total precision cutting company, possessing wire EDM, laser, and abrasive waterjet capabilities. I designed my new building with these objectives in mind. To examine abrasive waterjets, I flew to Flow International and visited various abrasive waterjet companies.

While planning my building, I bought a 1500 watt CO_2 laser. Even though it was a high-quality laser, I discovered lasers were a high maintenance item. Our company was so busy with wire EDM that I became the primary laser repairman. Again I decided to retreat and sell the laser.

After deciding to be just a specialized wire EDM job shop, I redesigned my building and enlarged our wire EDM department. Less than two years later, the enlarged EDM department was too small—we expanded to the other side of our building.

My seeming failures have become extremely valuable assets for me in writing this book. So whatever comes my way, I can count on the Lord to works things out for good. And I may add—it certainly has. Those having the Lord as their partner have a wonderful promise to rely on. It's a promise that brings me tremendous peace. The Bible says, "And we know that in all things God works for the good of those who love him, who have been called according to his purpose."[36]

Company and Personal Goals

Those of us in business should be willing to ask ourselves some challenging questions. What's my purpose for being in business? What are my personal goals? It's amazing how much time we can spend being busy for the material things of life, and take little time on taking inventory of our personal lives. Deep personal evaluation requires time and can be painful, but the wise will not flinch from making such decisions.

Perhaps by your shrewd business practices and investments you'll become a millionaire or a multi-millionaire; but remember, you can't take one cent of it with you when your time comes to leave this world. I like what one of my former pastors said, "You never see a U-haul following a hearse."

Solomon said, "The fear of the Lord is the beginning of knowledge."[37] Solomon's father, King David, said, "The fool says in his heart, 'There is no God.'"[38] If there's no God, what is life's purpose? Serving the Lord puts purpose and meaning into life. I made that discovery when I was 19 years old. I had a definite conversion experience where I repented of my sins and accepted Christ as my Lord and Savior. That decision radically altered my life. I wouldn't exchange anything in this world for that experience. It gave me peace and purpose for living.

My dream, from the time I started Reliable EDM, was to publish books to help people. In order to accomplish this, I needed money. The success of our EDM business was vital to the success of my publishing dreams. So as the needs arose, I bought more wire EDM machines. Then I put up an 18,800 square foot building which would provide room for both our EDM and publishing company. In addition, we built a digital recording studio and high-quality video studio in one portion of the building. One of the reasons for the video studio was to make teaching videos for the *Number Success* math series that I wrote. For years, my sons, Steve and Phil, managed the EDM business as I wrote books, including this one.

Often I ask people, "What constitutes value one hundred years from now?" Often they stand there and ponder that question. The answer? The only thing that constitutes value is what we've done for the Lord. One hundred years from now, your chances of being alive are extremely slim. The Bible tells us that every living person will spend eternity in either heaven or hell. If that is the case, then nothing is more important than making sure that when your life is over, you'll spent eternity with the Lord. Jesus said, "Do not store up for yourselves treasures on earth, where moth and rust destroy, and where thieves break in and steal. But store up for yourselves treasures in heaven, where moth and rust do not destroy, and where thieves do not break in and steal. For where your treasure is, there your heart will be also."[39]

If you do become successful, make sure that the Lord is in your life. For true success is having the smile of God. Solomon offered this wise advice, "Do not wear yourself out to get rich; have the wisdom to show restraint. Cast but a glance at riches, and they are gone, for they will surely sprout wings and fly off to the sky like an eagle."[40]

Listed are some benefits Solomon predicted would follow those who honor the Lord:

Honor the LORD with your wealth, with the firstfruits of all your crops; then your barns will be filled to overflowing, and your vats will brim over with new wine.[41]

One man gives freely, yet gains even more; another withholds unduly, but comes to poverty. A generous man will prosper; he who refreshes others will himself be refreshed."[42]

Making the Lord your CEO and blessing others is the wisest thing you can do. Seeking His approval and following the Lord's commandments will bring success. Solomon stated, "Commit to the LORD whatever you do, and your plans will succeed." [43]

Facing the World of Tomorrow

Facing the world of tomorrow with the Lord as CEO and having a burning desire to bless people provides us all with an exciting future. With this as our goal, we can go into the world of manufacturing to learn and implement the latest technologies so we can build a successful company and please the Lord.

Let us learn about the new developments that manufacturers are constantly introducing so we can build a successful business. Our generation has witnessed an explosion of new ideas and technologies.

Some machines come equipped with speedier 32-bit controllers and with fuzzy logic that allows machines to alter power according to various machining conditions. Drawings are sent on the Internet or by modem to other computers. Programs are available with automatic nesting capabilities. Computer programmers are constantly developing CAD/CAM (Computer Aided Design/Computer Aided Manufacturing) that improves the design and manufacturing of 3D shapes. Inspection equipment is available to automatically compare computer drawn parts with manufactured parts to reveal whether parts are in or out of tolerance. Artificial intelligence allows machines to think through complex machining sequences. Virtual prototyping and manufacturing allows engineers to make a three-dimensional model directly from their computers and to make necessary changes before employing hard tooling.

We have examined wire EDM, ram EDM, fast hole EDM drilling, abrasive flow machining, ultrasonic machining, photochemical machining, electrochemical machining, plasma and precision plasma cutting, waterjet and abrasive waterjet machining, various laser systems, and rapid prototyping and manufacturing. Those implementing successful business principles, and staying abreast of and applying wisely the new technologies to satisfy customers' demands, will forge ahead to become tomorrow's manufacturing leaders.

I would like to close this chapter with two famous quotes from the Apostle Paul.

Paul was a dedicated Jewish leader who accepted Christ after he had a dramatic encounter with him as he was traveling to Damascus. Paul labored extensively to tell both Jews and Gentiles about the good news of Jesus Christ. In a letter to the Romans he wrote, "Therefore, I urge you, brothers, in view of God's mercy, to offer your bodies as living sacrifices, holy and pleasing to God—this is your spiritual act of worship. Do not conform any longer to the pattern of this world, but be transformed by the renewing of your mind. Then you will be able to test and approve what God's will is—his good, pleasing and perfect will." [44]

Paul lived this way, and at the end of his life from a Roman prison he wrote these triumphant words shortly before his execution, "I have fought the good fight, I have finished the race, I have kept the faith. Now there is in store for me the crown of righteousness, which the Lord, the righteous Judge, will award to me on that day—and not only to me, but also to all who have longed for his appearing."[45] That's success!

References

1. Matthew 7:12
2. Matthew 22:35-40
3. Proverbs 13:11
4. 1 Kings 3:7-14
5. Proverbs 26:10
6. Proverbs 8:10-11
7. Proverbs 18:15
8. Proverbs 23:23
9. Proverbs 12:15
10. Proverbs 15:22
11. Proverbs 19:20
12. Proverbs 20:5
13. Proverbs 26:12
14. Proverbs 28:26
15. Proverbs 10:4
16. Proverbs 12:11
17. Proverbs 14:23
18. Proverbs 14:15
19. Proverbs 29:20
20. Proverbs 15:12
21. Proverbs 15:31-32
22. Proverbs 22:29
23. Proverbs 27:23-24
24. Proverbs 3:11-12
25. Proverbs 12:1
26. Proverbs 13:18
27. Proverbs 22:1
28. Proverbs 24:27
29. Proverbs 16:8
30. Proverbs 21:6
31. Proverbs 11:1
32. Matthew 25:21
33. Proverbs 24:19-20
34. Proverbs 6:6-8
35. Proverbs 24:10
36. Romans 8:28
37. Proverbs 1:7
38. Psalms 14:1
39. Matthew 6:19-21
40. Proverbs 23:4-5
41. Proverbs 3:9-10
42. Proverbs 11:24-25
43. Proverbs 16:3
44. Romans 12:1-2
45. 2 Timothy 4:7-8

Unit 15

The Revolutionary Future Non-Traditional Machine

Notes

30 The Revolutionary Future Non-Traditional Machine

How would you like to become a multi-millionaire or even a billionaire? You can if you can develop a non-traditional machine as will be described. This chapter is dedicated to those visionaries willing to pierce the unknown and use their talents and energy to develop this revolutionary non-traditional machine.

The Problem

Currently, the most accurate system for cutting thick materials is wire EDM. No non-traditional cutting system comes close to the accuracies achieved by this method. At present, the wire EDM process competes in many areas of traditional machining; however, compared to abrasive water jet, precision plasma, and lasers, wire EDM cuts slower.

The problem with abrasive water jet, precision plasma, and lasers is that the applied energy of these cutting systems tend to diffuse when cutting thick materials. This diffusion of energy creates undesirable finishes and accuracies.

At present six energy sources are used for cutting: electricity, water, abrasives, chemicals, light, and plasma. No system has yet been developed with these energy sources that can produce a straight cut on thick parts without the means of an electrode.

Possible Solutions

As stated previously, energy diffusion causes undesirable surface finishes. Therefore, a possible solution is to create a system which will supply sufficient energy which will not diffuse as it cuts thick material.

Such a system can create a high-energy ray beam with such an intensity that when it strikes any type of material, conductive or non conductive, it will pierce and not diffuse. If such a system were developed, it would cut materials with speeds that would revolutionize many industries.

What can be done to create an intensifier with such a high-energy beam? Here are some ideas:

A. Develop an intensifier that will use a magnetic field to spin and accelerate electrons with plasma.

B. Use a magnetic field with high-intensity laser light.

C. Have a high-energy magnetic field to guide plasma or laser.

D. Create a powerful laser wave length that will not diffuse.

E. Combine high-energy light, plasma, and electricity.

F. Use controlled atomic energy.

G. Apply star wars technology for cutting.

Conclusion

Imagine such a machine with a super high-energy intensifier cutting materials through 3 inches (76 mm) of hardened steel at speeds of 30 inches per minute with wire EDM finishes. Such a machine would make nearly all present cutting systems obsolete.

People have often laughed at dreamers, saying, "It can't be done."

But today our world is a much better place to live because there were those who let their minds dream and refused to accept the status quo. To the visionaries I leave this challenge: Make this world a better place to live by using your ingenuity to develop this revolutionary non-traditional cutting machine.

Unit 16

Questions*

*Permission is granted to teachers to make copies of these questions for student use.

Notes

31 Questions

Chapter 1
Fundamentals of Non-Traditional Machining

1. Why is the term non-traditional machining really a misnomer?

2. List five items that have revolutionized machining since the industrial revolution.

3. What is fuzzy logic?

4. What is virtual reality?

5. What does virtual prototyping allow engineers to do?

6. Identify and describe the two phases of the computer revolution.

7. List the six processes and the machines that are used in non-traditional machining. Also describe how each process alters material.

8. Why is it so important to understand the speed, accuracy, and characteristics of these various non-traditional machining methods?

9. What is one of the biggest difficulties concerning accuracies in the machining trade? Why is this issue so important?

10. If a ten inch piece of steel heats up ten degrees, how much will it expand?

11. What is one of the most critical issues for companies to remain successful?

Chapter 2
Fundamentals of Wire EDM

1. How have wire EDM cutting speeds changed since wire EDM was introduced?

2. How accurately can wire EDM cut?

3. What are design engineers doing as they discover the advantages of wire EDM?

4. What is the difference in speed between cutting exotic alloys and mild steel using wire EDM?

5. What is deionized water, and what does it do?

6. What happens between the electrode and the workpiece when sufficient voltage is applied?

7. What is the function of the pressurized dielectric fluid?

8. What is the function of the servo system?

9. Describe the step by step EDM process.

10. List the three types of wire EDMs. Describe each system.

Chapter 3
Profiting With Wire EDM

1. What are manufacturers discovering about the use of the new generation of high-speed wire EDMs?

2. Describe the accuracies and finishes achieved with wire EDM.

3. Draw a picture of an edge that has been stamped and an edge that has been wire EDMed.

4. Describe how damaged parts can be repaired with wire EDM.

5. Why is there decreasing need for skilled craftspersons?

6. What affect does material workpiece hardness have with wire EDM?

7. What is digitizing?

8. Why is wire EDM so reliable?

9. List ten parts that can be made with wire EDM.

10. What advantage is it to cut thin shims with wire EDM rather than with laser?

11. What type of material can be cut with wire EDM?

12. How are machining costs determined for wire EDM?

Chapter 4
Proper Procedures

1. In planning work for wire EDM, what is a good way to visualize the machine?

2. List the three methods to pick up dimensions on a part.

3. What is automatic pick up?

4. What happens if the holes are not square when they are picked up?

5. Where is the best place to put starter holes?

6. Where should starter holes be placed for cutting out thin slots? Why?

Chapter 5
Understanding the Wire EDM Process

1. What tolerances can wire EDM machines hold?

2. Why is wire EDM able to get such a fine finish on even tall parts?

3. What is the wire kerf for a .012" (.030 mm) wire?

4. What will always happen when inside corner radii are machined with wire EDM?

5. What must be done to achieve very sharp outside corners?

6. List the three main reasons for skim cuts.

7. What determines the hardness and toughness of tungsten carbide?

8. When tungsten carbide is EDMed, what is eroded away during the EDM process?

9. What is polycrystalline diamond?

10. What two things does the pressurized deionized fluid do?

11. What precautions can be taken to avoid flushing pressure loss?

12. What is the wire EDM ideal metal thickness to obtain the maximum square inches of cut per hour?

13. List some outcomes when the wire electrode meets impurities.

14. Describe recast and heat-affected zone.

15. What practically eliminates the heat-affected zone?

16. What do some wire EDM machines come equipped with to minimize heat-affected zones?

17. Describe and list the advantages of non-electrolysis power supplies.

18. What is the advantage of heat treating steel before the EDM process?

19. When EDMing large sections, list the actions that can be taken to relieve inherent stresses.

20. List the reasons to leave a frame around the workpiece.

21. Why is it important that on some operations the frame should have sufficient strength?

Chapter 6
Reducing Wire EDM Costs

1. Why does creating one slug with wire EDM reduce costs?

2. Why is having the flush ports on the workpiece the most efficient way to cut for wire EDMing?

3. Describe air machining.

4. Give an example of when it is better to machine parts after they have been EDMed.

5. When parts are stacked to be EDMed, what must be avoided?

6. When production lots and multiple parts are stacked together, what must be considered to make the job most efficient?

7. Why does cutting with thin wire electrodes increase costs?

Chapter 7
Various Wire EDM Applications

1. What is the predominant type of work at the author's wire EDM job shop?

2. What is one of the greatest problems with designers and engineers concerning wire EDM?

3. How did the author solve the problem with designers and engineers?

4. Pick four cutting operations and describe each one.

5. Describe the old way of making tool and dies. (Before precision grinding.)

6. A. Describe how precision tools and dies were made before wire EDM.

   B. Describe the skill that was required to make tool and dies with precision grinding. (Study the author's handwritten notes.)

7. A. Describe how tools and dies are made using wire EDM.

   B. What effect has wire EDM had on tool and die makers?

8. What has been the overall effect of wire EDM on tool and die making?

9. List the advantages of wire EDMed dies.

Chapter 8
Fundamentals of Ram EDM

1. List the various names for ram EDM.

2. What is ram EDM generally used for?

3. List and explain the two significant improvements in spark erosion from the two Russian scientists.

4. How did the transistor aid in ram EDM?

5. What does the DC power supply provide?

6. What function does the dielectric oil have when electricity is first supplied?

7. What happens when sufficient electricity is supplied between the electrode and the workpiece?

8. What happens during the off time of the electrical cycle?

9. What effect does polarity have on the workpiece and the electrode?

10. What happens in the no wear cycle?

11. Describe fuzzy logic.

12. What should be done concerning fumes from ram EDM?

Chapter 9
Profiting With Ram EDM

1. When is it profitable to machine blind cavities with ram EDM?

2. Why is it possible to EDM thin sections?

3. What is the possible accuracy of ram EDM?

4. List two methods of putting threads into hardened parts.

5. List four applications for ram EDM.

6. What kinds of materials can be machined with ram EDM?

7. How can mold makers increase the speed of their molds?

8. Why does carbide cut slower than steel?

Chapter 10
Ram EDM Electrodes and Finishing

1. What is the function of the electrode?

2. List the factors that need to be considered in selecting the electrode material.

3. What are the two main types of electrode material?

4. Why is graphite a commonly used electrode material?

5. Describe the Galvano process for metallic electrodes.

6. How are custom molded metallic electrodes made?

7. What is one of the major problems with graphite?

8. What factors determine the cost and cutting efficiency of graphite?

9. What are the two general rules for choosing the type of graphite material?

10. Where does the heaviest electrode wear occur? Why?

11. Describe the process for abrading graphite electrodes.

12. Describe the ultrasonic machining process for graphite electrodes.

13. What is an efficient way to machine metallic and graphite electrodes?

14. What determines the amount of overcut that occurs in an EDMed cavity?

15. When do maximum and minimum overcuts occur?
    Explain the reasons.

16. List the four types of surfaces that occur when EDMing.

17. Why can there be significant differences in the heat-affected zones between wire and ram EDM?

18. What effect does the dielectric oil have on the recast surface?

19. What has happened with the newer power supplies concerning heat-affected zones?

20. What significantly reduces heat-affected zones?

21. What causes rough and fine finishes when EDMing?

22. What have some manufacturers done to produce mirror finishes?

23. Describe the micro machining process.

Chapter 11
Dielectric Oil and Flushing for Ram EDM

1. Describe the three important functions of the dielectric oil.

2. What is flash point?

3. List some of the factors that make flushing so important?

4. What happens with improper flushing?

5. What happens when arcing occurs?

6.   When and why is arcing most likely to occur?

7.   What is important concerning the volume and pressure of the dielectric oil?

8.   List and describe the four types of flushing.

9.   List and describe the three types of pulse flushing.

10. What does the filtration system do?

Chapter 12
Reducing Costs for Ram EDM

1.   In machining large cavities, what helps to reduces costs?

2.   Describe the different procedures for cutting a hex with ram and with wire EDM.

3.   With the advent of solid-state power supplies and premium electrodes, what is now possible with roughing electrodes?

4.   What are the advantages of electrode and workpiece holding devices?

5.   How has orbiting reduced costs concerning electrodes?

6.   How does the orbital path aid in flushing?

7.   In orbiting, both the bottom and the sides of the electrode can be used for finishing. How does this reduce costs?

8.   List the twelve possibilities for orbiting machining.

9.   Describe microwelding.

10. Describe the use of abrasive flow machining to remove recast layer from EDMing.

11. Describe the use of tool changers.

12. What is hard die milling?

13. What are the possibilities with hard die milling?

Chapter 13
Fast Hole EDM Drilling

1. What kind of materials can fast hole EDM drill?

2. List some applications for fast hole EDMing.

3. Describe how fast EDM drilling works.

4. What does the dielectric fluid accomplish?

5. What does the servo mechanism do?

6. What kind of electrodes are used for fast hole EDMing?

7. How deep have holes been drilled using this process?

8. What benefit does the high pressure have on the servo unit?

9. What is the major difference between a metal disintegrating machine and a fast hole EDM drill?

10. What happens to cutting speed when large holes need to be drilled with fast hole EDMing?

11. Compare drilling on curved and angled surfaces with conventional drills and fast hole EDMing.

12. What is the difference concerning the burrs from conventional drilling versus fast hole EDMing.

13. When are conventional drills most likely to break?

14. What is the difference in the torque conditions when drilling with conventional machines versus EDM?

15. Why does fast hole EDMing produce straight holes?

Chapter 14
Abrasive Flow, Thermal Energy Deburring, and Ultrasonic Machining

1. What is abrasive flow machining?

2. Describe abrasive flow machining process.

3. What must happen in the media flow area for abrasive flow machining to occur?

4. What is the basic rule for abrasive flow machining?

5. What determines the speed and finish of machined surfaces?

6. What differences occur between stiff or thinner media?

7. List the uses for abrasive flow machining.

8. Describe the process of thermal energy deburring.

9. What are the limitations of thermal energy deburring?

10. For what is ultrasonic machining particularly useful?

11. Describe the process of ultrasonic machining and polishing.

Chapter 15
Photochemical Machining

1. What are some of the other names of photochemical machining?

2. Why are tabs used in designing the part for photochemical machining?

3. What is applied to the surfaces to be etched?

4. After the material is coated, how is the image produced on the material?

5. Describe how the etching and stripping process creates the desired part.

6. List some of the users of photochemical machining.

7. What are the general tolerances for this process?

8. Describe how cutting and etching can be done in one operation.

9. Describe how three-dimensional etching is done.

10. What kind of edges are always produced with photochemical machining?

11. What are the advantages and disadvantages of photochemical machining?

12. Why is length of cut not an important cost factor with photochemical machining?

12. What is the general rule for photochemical machining when creating holes for micro-etching screens?

13. What is the advantage of etched screens compared to woven screens?

14. Describe the process of electroforming micro-etched screens.

Chapter 16
Electrochemical Machining

1. To what non-traditional machining method is electrochemical machining similar?

2. Describe the electrochemical machining process.

3. How is the ECM process opposite to plating?

4. What is the difference between the shape of an electrode using ram EDM and an electrode using ECM?

5. Why is specialized tooling required for ECM?

6. Why is sludge removal an important issue with ECM?

7. What is one of the major advantages of the ECM process concerning the electrode?

8. Why is there no recast or thermal stress with ECM?

9. Describe the stem drilling process.

10. What is the difference between stem drilling and capillary drilling?

11. What are the three major difficulties with ECM?

12. When is it profitable to use ECM?

Chapter 17
Plasma and Precision Plasma Cutting

1. What kind of gas is used for plasma cutting?

2. Explain the difference between a regular and a precision plasma cutting system.

3. What are the four states of matter?

4. What is plasma?

5. What happens to the gas as it flows through the nozzle?

6. Describe how plasma cuts.

7. Describe dual gas plasma cutting.

8. Describe water shield plasma cutting.

9. Describe water injection plasma cutting.

10. What makes precision plasma cut more accurately?

11. What are the two major problems with conventional plasma cutting?

12. How does precision plasma overcome these two major problems?

13. What are the general cutting thicknesses for conventional and precision plasma cutting machines?

14. What kinds of materials can plasma cut?

15. What kinds of materials can lasers cut?

16. What is the difference between heat absorption in cutting between conventional and precision plasma?

17. What factors are critical for cutting accuracy?

18. What are the differences between precision plasma and laser cutting machines?

19. A. Describe a plasma and turret punch press.

    B. What is the advantage of such a machine?

Chapter 18
Waterjet and Abrasive Waterjet Machining

1. Describe waterjet cutting process.

2. How does abrasive waterjet cutting work?

3. What thickness can abrasive waterjet cut?

4. What type of material is generally used for cutting?

5. List the four parts of the abrasive cutting head and their functions.

6. Why doesn't the motion system have to be massive?

7. Why is a catcher system needed with waterjet machining?

Chapter 19
Profiting With Waterjet and Abrasive Waterjet Cutting

1. List ten materials that can be cut with waterjet.

2. List ten materials that can be cut with abrasive waterjets.

3. What has a great effect on cutting accuracy with abrasive waterjets?

4. What happens to the bottom of thick cuts with abrasive waterjets?

5. How loud is the noise level on some machines, and how do some manufactures lower it?

6. List the four disadvantages for waterjet and abrasive waterjet cutting.

7. List six advantages for waterjet and abrasive waterjet cutting.

8. Describe the glass sculpturing process.

Chapter 20
Fundamentals of Lasers

1. Describe the power of coherent laser light beams.

2. What is one of the major reasons lasers are so cost effective?

3. What does the word "laser" stand for?

4. Describe how a laser light is created in the resonator.

5. Describe the differences between reflective mirrors and partially reflective mirrors.

6. What is the purpose of laser optics?

7. What determines what kind of lenses are used when cutting?

8. How does oxygen as an assist gas help in metal cutting?

9. A. What is done to provide an oxide-free edge for stainless steel?

   B. For what is this edge useful?

10. Describe the laser cutting process.

11. What is the purpose of the sensing unit?

12. Why do precautions need to taken with:

   A. Laser beams?

   B. Laser fumes?

Chapter 21
Understanding Laser Cutting

1. What is the general kerf width for cutting lasers?

2. What produces most of the distortion when cutting with lasers?

3. Why do lasers have a small heat-affected zone?

4. What are some of the factors that determine edge quality when materials are laser cut?

5. What happens to edge quality on thicker materials?

6. Why are test cuts generally needed for laser cutting?

7. What are the factors that determine the cutting speed of lasers?

8. How does material surface condition affect laser cutting?

9. To achieve maximum laser efficiency, what must be done with the laser beam?

10. At what place is the laser beam most efficient?

11. What are the advantages of laser cutting compared to using hard tooling?

12. What are the advantages of using lasers and turret punch presses?

13. What can turret punch presses do that lasers cannot?

Chapter 22
Various Lasers and Their Configurations

1.   What are the two lasers that dominate the material-processing field?

2.   Describe how Nd:YAG lasers work.

3.   What is done to Nd:YAG lasers to increase their power?

4.   What determines the different kinds of lasers?

5.   Why is fiber optics such a benefit for Nd:YAG lasers?

6.   How long a cable can be used with fiber optics?

7.   What is timesharing of the laser beam?

8.   Why is Nd:YAG laser an ideal machine for drilling compared to CO_2 lasers?

9.   What is the niche for excimer lasers?

10. Describe how excimer lasers differ from other lasers.

11. What kind of heat-affected zone do excimer lasers have? Why?

12. A. How small a hole have excimer lasers been able to machine?

     B. If a human hair is .002" (.05mm), how many side by side holes can be put across it?

13. Describe the system of masking with excimer lasers.

14. Describe how integrated circuits are made with optical microlithography.

15.   What is the hindering factor in producing smaller integrated circuits?

16. List the three basic traveling methods for lasers and describe each.

17. Explain how beam splitting works.

Chapter 23
Profiting With Laser Cutting

1.   List at least ten materials that lasers can cut.

2. What two considerations need to be made concerning cutting various materials?

3. What kinds of materials do lasers have trouble cutting? Why?

4. A. How thick can a 1500 CO_2 laser cut carbon steel?

    B. How thick can a 3000 CO_2 laser cut carbon steel?

5. Describe how tube cutting lasers work.

6. What can multi-axis lasers do?

7. What does a sheet loader do?

8. What is time sharing?

9. How fast can some lasers switch in time sharing?

10. Why are there substantial savings on material costs with lasers?

11. Why do lasers produce a small heat-affected zone?

12. Why are lasers able to cut thin webs?

10. Explain why lasers are so useful for JIT machining.

11. Explain the disadvantages of lasers concerning:

    A. Costs.

    B. Maintenance.

    C. Optics.

Chapter 24
Lasers for Welding, Drilling, Cladding, Alloying, Heat Treating, Marking, and Drilling

1. What are the two common lasers for welding?

2. When does welding occur?

3. What kind of parts are good prospects for laser welding?

4. What is the difference between laser welding and electron beam welding?

5. Describe the two types of laser welding.

6. How do lasers weld steel and copper together?

7. Describe keyhole welding.

8. What is the difference between conventional spot welding and laser welding?

9. Describe the process of laser cladding.

10. What is hardfacing?

11. What are some of the difficulties associated with conventional cladding methods?

12. What are the advantages of laser cladding?

13. Describe laser surface alloying.

14. Why are lasers an ideal tool for surface alloying?

15. What is the difference between laser alloying and laser cladding?

16. Describe laser heat treating and its process.

17. How is the round laser beam converted to a square beam?

18. What does the laser beam do to the surface of the metal in heat treating?

19. What kind of metals are used for laser hardening?

20. What determines the depth of hardness?

21. List the advantages and disadvantages of laser hardening.

22. List some of the varied products that are useful for laser marking.

23. Describe the three methods for surface alterations used with laser marking.

24. Explain the difference between mask laser marking and beam-steered laser marking.

25. What is the major advantage of laser marking?

26. What is another advantage for laser marking compared to ink or dye printing, or chemical etching?

27. How fast can some lasers drill holes?

28. What are the general limitations for laser drilled holes?

29. Describe how laser drilling works.

30. List some of the disadvantages with laser drilling.

31. List some of the advantages with laser drilling.

32. List some of the exciting developments with lasers.

Chapter 25
Rapid Prototyping and Manufacturing

1. What are some of the names for rapid prototyping?

2. What is virtual reality?

3. What does rapid prototyping allow engineers and designers to do?

4. Define CAE/CAD/CAM.

5. Describe the computer aided design process for the photopolymer rapid prototyping system.

6. Describe the process of photopolymerization.

7. What does the translator do?

8. Describe the sweep and Z-wait.

9. How is the final product cured?

10. What can be done with the product after it has been cured?

11. How are larger parts made that go beyond the envelope of the machine?

12. Describe the laminated object prototyping process.

13. What kind of material is used in the laminated object prototyping process?

14. What is a similar material?

15. Describe the selective laser sintering process.

16. What kinds of materials can be used with laser sintering?

17. What can be done with patterns produced in wax?

18. Explain how the metal powder composed of steel/copper and thermoplastic is made into a solid metal part.

19. Describe the 3D Keltool™ process.

20. Describe the solid ground curing process.

21. After the model is completed in the solid ground curing process, what is the model encased in?

22. What does solid ground curing not need because it is done in a solid environment?

23. Explain the fused deposition prototyping process.

24. What is the function of the extrusion head in fused deposition prototyping?

25. What kinds of materials can be used with fused deposition prototyping?

26. Describe the ballistic particle manufacturing process.

27. How does the jetting system work in the ballistic particle manufacturing process?

28. Describe the multi-jet modeling process.

29. What is multi-jet modeling process system similar to?

30. Describe the ProMetal process.

31. What is done with the "green form" in the ProMetal process to make it a solid metal part?

32. What is the great advantage of office modeling systems?

Chapter 26
Understanding the Rapid Prototyping and Manufacturing Process

1. How is rapid prototyping and manufacturing changing the way companies are doing business?

2. How does rapid prototyping eliminate time-consuming and costly model making?

3. In the earlier years of rapid prototyping it was used primarily for reviewing prototypes. How is it now also being used?

4. How does rapid prototyping enhance design verification?

5. Describe how rapid prototyping is used with investment casting.

6. Describe rapid milling for prototypes.

7. Since many changes are taking place in this field, what should manufacturers do to make sure they have the best system?

Chapter 27
Purchasing the Right Equipment

1. What are two difficult decisions that confront manufacturers concerning the purchase of equipment?

2. What is one of the major advantages of having equipment on premises?

3. List questions that need to be answered on purchasing a piece of equipment concerning the cutting of holes.

4. List processes that can cut holes.

5. List processes that can cut flat blanks.

6. Why is the condition of the motion system important?

7. What does wire EDM do that makes it so accurate?

8. Why is choosing the right brand of a machine so important?

9. Give reasons why the following are important before choosing to buy a machine:

   A. Service.

   B. Capability.

   C. Reliability of Company.

10. List at least five factors that should be considered when purchasing equipment.

11. What kind of environment needs to be provided for close tolerance work?

Chapter 28
Benefits of Using Contract Shops

1. What is often the mentality of workers in a:

   A. Manufacturing plant?

   B. Job shop?

2. List and describe the four advantages of using job shops.

3. What is another advantage of job shops concerning fixtures and machine modifications?

4. What kind of personnel are needed to operate non-traditional machines? Explain why.

5. Take a wire EDM machine that costs $200,000. Estimate the cost for:

   A. Loan interest for five years.

   B. Programming and inspecting equipment.

   C. Training personnel

   D. Additional space.

   E. Machine repairs and maintenance.

   F. Machine and operation costs.

8. A. Pick a non-traditional machining machine you are interested in, and fill out the questionnaire. Estimate the amount for each cost. Permission is granted to make copies of the "Five Year Estimated Cost."

    B. What is the yearly cost for your machine?

9. Before making the final decision to purchase a non-traditional machine, what should be considered?

10. Explain the concept: maximizing core competency.

11. What have many successful companies discovered?

Chapter 29
Building a Successful Business

1. A. What was the total non-farm employment in 1950 in manufacturing?

    B. What is it today?

2. What is the major reason for this decline?

3. A. How many were engaged in agriculture in the beginning of the 19th century?

    B. How many are today?

4. What can we learn from these trends?

5. What is the cornerstone for raising a nation's standard of living?

6. What can we learn from the author's experience before he started Reliable EDM?

7. State the author's philosophy of success.

8. What are the two foundational truths the author discovered for building a successful business? Explain each.

9. Define quality.

10. Where is the best place to insure quality?

11. Define service.

12. What did the author do when he became overloaded with work?

13. What is commonly said about quality, service, and low prices?

14. What did the author do in his business to lower costs?

15. What is the most important principle for building a successful company?

16. What should a business teach its customers?

17. What is one of the practices of successful stores?

18. What did King Solomon ask for?

19. What did King Solomon receive?

20. List the things that the author does to hire qualified workers.

21. Pick an occupation you are interested in pursuing. If you were the one doing the hiring, list the things you would desire for the applicant.

22. If you plan on going into business, list some things you should you do before you start your business.

23. List some things business owners can do to avoid workers from wasting time in searching for tools.

24. In seeking advice, what did the author do in regards to this book on non-traditional machining?

25. How can you obtain valuable information from others?

26. What is trying to achieve success without doing hard work?

27. A. What does a simple man do?

    B. What does a prudent man do?

28. A. What will an open minded individual do?

    B. What test does the author give to determine an individual's open-mindedness?

29. What kind of experience should a person have if he wants to start a business?

30. You have reached the pinnacle of success. What must be done continually to remain successful?

31. What are economic forces constantly trying to do?

32. Define a permissive approach in doing business.

33. What should one do when they encounter problems in the workplace?

34. What did the author determine to do when he began his company?

35. What did the author do to make sure he had the latest and best technology?

36. What is a company's best advertisement?

37. What steps can be taken for start-up companies to establish credibility?

38. What is the purpose of the author of having a display in the foyer?

39. When profits are made, what should take priority?

40. What is one of the fastest ways to destroy a company?

41. What precautions should be taken about going heavily in debt?

42. What is the best way to avoid having troublesome workers?

43. If a portion of the business is unprofitable, what should be done?

44. What will wise individuals do concerning their company and personal goals?

45. What did King Solomon say about riches?

46. List some of the new developments that are being introduced into the market.

Chapter 30
The Revolutionary Future Non-Traditional Machine

1. What is the problem cutting thick parts with abrasive water jet, precision plasma and lasers?

2. List the six energy sources that are being used to cut thick materials.

3. At present, what must be used to produce a straight cut on thick parts?

4. If energy diffusion is the basic problem in machining, what is a possible solution?

5. Describe at least three possible solutions for developing this machine.

Index

Notes

Notes

Notes

Reading Success
Phonics-Literature Reading Program

Success in today's world requires an ability to read. *Reading Success* is unique in that it teaches students not only to learn new words, but combines an intensive phonics program with action-oriented stories. Knowledge learned is immediately applied by reading interesting material.

Reading Success can be used in the classroom or used with audio as a self-directed program. After students learn the letter sounds, they are taught words by blending the various sound patterns. After mastering these sounds, students immediately put into practice what they have learned by reading exciting stories. Workbooks reinforce the reading and teach grammar.

See for yourself why this reading program has received such high acclaims. Examine the material for thirty days, and if for any reason you're dissatisfied, simply return them for a full refund.

Reading Success for Adults Textbook and Workbook 1

Textbook Contents: Students are first taught the letter sounds of the alphabet. Then they learn to blend words having the short vowels: a e i o u. In the each of the lesson books interesting stories are read using the words students have learned.

Workbook Contents: Students answer questions for textbook and readers.

Readers: Adventures and Surprises
After students learn the short vowel words in each lesson, then they read stories first from *Adventures* and then from *Surprises*.

Reading Success for Adults Textbook and Workbook 2

Contents: Combination sounds: th ch sh and wh. Making words plural by adding "s." Silent e. Long vowels a e i o u. Letters "y" and "w" as vowels. Short vowel s a e i o u.

Workbook Contents: Students answer questions for textbook and reader.

Readers: Victories
The Big, Big Fish; The Hot Rod; and Mountain Climbing.

Reading Success for Adults Textbook and Workbook 3

Contents: Long vowels a e i o u. Twenty nine consonant blends. Compound words. Two to six syllable words. Making words plural. Rule breakers and non-rule breakers. Sound patterns: th ck ph tch.

Workbook Contents: Students answer questions for textbook and reader, and learn grammar.

Readers: Struggles
Your Job is Easy; Two Tears; and No Longer a Dilly Dally

Reading Success for Adults Textbook and Workbook 4

Contents: Silent letters: b g h k n t p w. Hard and soft c and g. Numbers, days and months. Contractions, rhyming words, and homonyms. Forty two suffixes and prefixes. Diphthongs ou and ough and various vowel combinations and sounds. Holidays and states. Commencement. Using a dictionary.

Workbook Contents: Students answer questions for textbook and reader, and learn grammar.

Readers: Future Trek and George Washington Carver: Making Much From Little

Reading Success for Adults: Supervisor's Instructions and Teaching Helps—Books 1-4

Contents: Books provide answers for supervisors or students to check their workbook answers. Instructions are provided for *Reading Success* and *Handwriting Success*. Also teaching aids are provided for teaching new words.

Handwriting Success for Adults
Learning Manuscript and Cursive Writing

Contents: Students can learn either manuscript or cursive writing. Many students when learning cursive writing are unable to read the words. By combining both types of writing, students are able to understand what they are writing.

Audio for Self-Directed Learning

Reading Success Endorsements

Reading Success is an excellent addition to the field of adult literacy. Its well developed, systematic approach will no doubt enable the adult non-reader to learn to read. The combination of phonics and whole language in one set of instructional materials is commendable.

The books are well designed with an appealing, sophisticated cover and appropriate fonts throughout the series. Incorporating instructor sections in the lesson itself is particularly helpful. There are regular review sections which are also well done.

Donna McCoy, *Program Specialist*

I liked the format of these books. There is plenty of practice of phonics skills as well as comprehension. I like the integration of grammar skills as the series advances. I also like the variety of stories and characters in the short story books. It gives new readers a taste for different types of literature and hopefully enough exposure so they can begin to make choices about reading.

Sandra Delgado, *Curriculum and Instructional Support Manager for Kentucky Department for Adult Education and Literacy*

I have been working with adult ESL and GED students for almost ten years. I think your materials will provide an excellent resource for the adult learner.

The stories presented in this series are very interesting and unique. The books offer phonics, vocabulary, and

grammar rules. The stories are ideal for my students because they are written for adults, not small children. The book about the life of George Washington Carver should provide a source of inspiration for underachieving students. The handwriting guide is one of the best guides that I have ever seen.

Your books are very good and if used within the structure of a well-balanced program, I have confidence that the books will greatly benefit all adult learners. Congratulations on an outstanding program.

William Dixon, *ESL and GED Teacher*

Through our Adult Education Program at North Harris College, our teachers work with Pre-GED, GED and Basic ESL students. I was very impressed with the quality of both the Reading Success and Number Success series....

Several teachers also looked at these materials and liked the straightforward approach of the books. Both the Reading Success and the Number Success have lessons based on reality built into them, which is something that our students usually need as well. The stories are engaging and the lessons very relevant to their everyday lives. The lessons are not too lengthy, so as not to lose the interest of the student.

The series will be a refreshing addition to classrooms of all ages everywhere.

Tracy Hendrix, *Adult Education Counselor, North Harris College*

For information visit our Web site: AdvancePublishing.com or call 800-917-9630

Number Success
Practical Problem Solving

Success in today's world requires an understanding of mathematics. *Number Success* is a series of adult math books that provide students with easy-to-learn workbooks so they can function intelligently in everyday living and in the workplace.

Along with the workbooks, videos supplement the lessons. The videos have experienced teachers, while Carl Sommer, author of *Non-Traditional*

Machining Handbook, goes to the field and demonstrates the many practical applications of math in machine shops, construction sites, stores, warehouses, and many other places.

Examine the material for thirty days, and if you don't believe that this is the best math program available, simply return the material for a full refund.

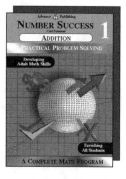

1) Addition

- Adding various place numbers
- Solving measurements
- Conversion Problems
- Adding money and problem solving

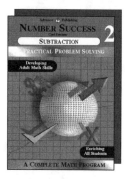

2) Subtraction

- Subtracting various place numbers
- Subtraction with borrowing
- Problem solving with addition and subtraction
- Check writing and bank deposits
- Comparing prices

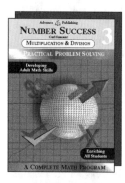

3) Multiplying and Dividing

- Multiplying and dividing various place numbers
- Comparing multiplication with division
- Multiplication and division problem solving
- Addition, subtraction, multiplication and division problems

4) Fractions

- Understanding and comparing fractions
- Writing equivalent fractions
- Raising and reducing fractions
- Adding, subtracting, multiplying and dividing various fraction problems
- Rulers and measurements

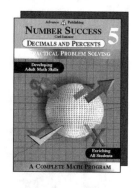

5) Decimals and Percents

- Comparing fractions, decimals and percents
- Adding, subtraction, multiplying and dividing decimals
- Changing fractions, decimals and percents
- Fraction, tenth, hundredth and metric rulers

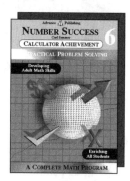

6) Calculator Achievement

- Calculator operation instructions
- Adding, subtracting, multiplying and dividing
- Weights, measures and interest problems
- Converting English and metric weights and measurements
- Solving fraction, decimal and percentage problems

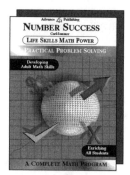

7) Life Skills Math Power

- Area and perimeter of rectangles, squares, triangles and circles
- Volume of rectangular solids, cubes, and cylinders
- Converting measurements
- Finding averages
- Many practical Life-Skills Math problems

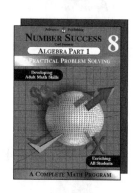

8) Algebra Part 1

- Adding, subtracting, multiplying, and dividing signed numbers and equations
- Practical applications for signed numbers, powers, and square roots
- Problem solving using formulas
- Application of algebraic expressions using formulas

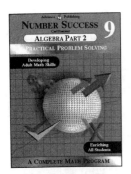

9) Algebra Part 2

- Solving equations with multiple unknowns
- Adding, subtracting, multiplying, and dividing algebraic terms
- Equations with symbols of inclusion
- Solving quadratic equations by factoring
- Consecutive integer problems

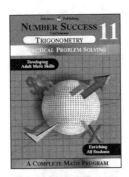

11) Trigonometry

- Formulas and solutions for finding angles and sides of right and oblique-angled triangles
- Ratio and angleYYs and sides of triangles
- Using a calculator for finding angles and sides of right and oblique-angled triangles
- Practical problem solving for right and oblique-angled triangles

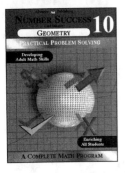

10) Geometry

- Formulas and Geometric symbols
- Angle problem solving
- Perimeter and areas of triangles and quadrilaterals
- Pythagorean theorem
- Practical geometric problem solving

12) Supervisor's Guide and Replacement Exams

- Instructions for supervisors
- Instructions for students
- Placement evaluation for books 1 to 7
- Placement evaluation for books 8 to 11

Videos for Self-Directed Learning

Number Success Endorsements

Number Success mathematics series is a self-paced comprehensive program which provides many opportunities to reinforce math skills....The program is very flexible and allows for various implementation models, such as, classroom, individual, or group instruction.

After reviewing and completing every math problem in the entire course, I highly recommend this program. It is an excellent program, and I look forward to using it when it becomes available.

Martha Froebel, *Director of Instruction, Designer of Math Programs, College of Education Math Methods Instructor, and Regional Mathematics Collaborative Consultant.*

In the math book series I was pleased to see there was practice for the students, and that students didn't learn a concept with only three problems. There was multiple practice with a lot of practical applications. This made the books very user-friendly for the students.

I think you have a winner with the math series.

June Vander Molen, *Educational Cooperative Program Specialist*

I like the step by step samples, explanations and the visual explanations. The way place value was explained is one of the easiest and clearest that I have seen. The math problems are very good practical adult situation problems. I like the practical problems but really like that information is provided in order to do the problems.

Sandra Delgado, *Curriculum and Instructional Support Manager for Kentucky Department for Adult Education and Literacy*

The books have very good explanations of the operations and concepts of materials that it covers. The visuals are especially helpful for students to understand the meaning of math concepts. There is good sequencing and building upon skills. It is stressed throughout all of the books that [it] is important to be neat in working math problems. The workbooks show how to neatly and methodically line up numbers for operation, both simple and more complex....I definitely recommend the math series.

Susan Stauber, *Teacher for Adult and Community Education in Tallahassee, Florida*

For information visit our Web site: AdvancePublishing.com or call 800-917-9630

Unique Self-Directed Literacy Programs for the Workplace

Number Success

A Practical Problem-Solving Adult Math Program With Videos

Number Success shows many practical examples of the application of math. There are twelve books to this series:

1. Addition
2. Subtraction
3. Multiplication and Division
4. Fractions.
5. Decimals and Percentage
6. Calculator Achievement

7. Life Skills Math Power
8. Algebra Part 1
9. Algebra Part 2
10 Geometry
11. Trigonometry
12. Supervisorís Guide

Reading Success

A Total Phonics-Literature Based Reading Program With Audio

Reading Success is an adult reading program that combines intensive phonics with action-packed stories. There are four lesson books, four workbooks, six illustrated readers, and four books for supervision and providing answers. There is also a workbook, *Writing Success*, which teaches manuscript and cursive writing.

Both programs have received high endorsements from those teaching literacy. See pages 429 to 433 for more details.

Advance Publishing, Inc. 6950 Fulton St., Houston, TX 77022 Ph. 713-695-0600 or 800-917-9630
Visit our Web site: **www.AdvancePublishing.com**